THE
PLACE-NAMES
OF
HUDDERSFIELD

First Published in 2008 by
David Shore, Huddersfield

ISBN: 978-0-9508526-9-0

Front cover illustration:
A view along the canal bank, with Brierley's Mill and
Turnbridge in the distance, c.1980.

Back cover illustration:
Cecil Street was already well developed by the 1870s and
is likely to have been named in honour of the Conservative
statesman, Robert Cecil, later to be Prime Minister. Back Cecil
Street, like other 'Back' streets, was originally a service road,
giving access to ashpits and privies.

Cover and text origination by
Greenwood (Recycled) Printing.
01484 844841

Printed in Great Britain by the MPG Books Group,
Bodmin and King's Lynn

CONTENTS

MAPS

ACKNOWLEDGEMENTS

The information in this work of reference comes from a wide range of sources and individuals, and my debt in those respects is acknowledged in the note on sources and in the text. Of scholars who have made a major contribution to the subject I would single out Edward Law, whose work has always been of the highest standard. When he left Huddersfield to live in Ireland he left a gap in Huddersfield's local history that has not yet been filled.

A further debt that I am more than happy to acknowledge is the one that I owe the late Clifford Stephenson. He passed on to me his huge collection of local photographs and pictures, in the hope that they would serve to illustrate a book such as this. I am delighted that I have been able to use so many of them here, in part repayment for his kindness.

I must also, once again, thank my publisher David Shore who spares no effort in his search for accuracy. He has made several important contributions to the book and saved my blushes more than once by identifying errors of fact and presentation. For example, the items on Bulstrode and Bradley Hall owe much to his vigilance and work in the field. If this book is more attractive than my early self-published works, and I am sure that it is, the credit is entirely his.

Finally, I should like to pay tribute to my wife Ann-marie for her patience and expertise. The computer is undoubtedly a great boon to writers but my relationship with the one that we possess is an uneasy one. Time and again she has rescued my text from oblivion and for that I am truly grateful.

PREFACE

There are two main aims in this dictionary. The first is to explain the origins of all the major place-names in the township of Huddersfield, including those for Bradley which have previously been largely ignored. Where possible I go beyond the etymologies and examine the special circumstances in which each name originated. This information is part of the warp and woof of the township's history over the last nine hundred years and it throws light on a number of topics such as clearance, settlement, husbandry, communications and industrial practices. The second aim is to include information about the major street names of the town, even where this involves crossing township boundaries in more recent times.

One important issue raised in this book has to do with the first Anglian settlement of Huddersfield, established possibly as early as the 7th century. Under the heading for Fartown I make the suggestion that the original site may have been in that area, close to Bay Hall, and that the nucleus around the church, the heart of the modern town, may date back only to the late 11th century when Huddersfield first became a parish. My view is that the new Norman landlords, the de Lacys, chose to erect their church on the south side of the town brook and not the north, a site that would have advantages for the parish as a whole. If that were the case it would throw new light on some very old problems and the idea seems at least worthy of consideration. The suggestion could be thought of as contentious, particularly as it may never be possible to prove or disprove it, but it is not as outrageous as might at first appear. Huddersfield would have had a very small population at that time and it is not unlikely that the houses were widely dispersed. The church gave it a status it had not previously enjoyed.

INTRODUCTION

In many south-Pennine townships the settlement in the medieval period was scattered and there was no nucleated community or 'town'. In Huddersfield though there is evidence to suggest that the church had emerged as the focal point of a small town by the 14th century. It had no market but its local status is clear in the poll tax returns of 1379 which list the population a generation after the Black Death.

Some local historians emphasise how unimportant Huddersfield was in comparison with Almondbury. After all Almondbury had market rights and fairs, and there had been an early attempt to establish it as a borough. It also lay in the shadow of Castle Hill which had been the site of an administrative centre for the Honour of Pontefract. Nevertheless, the 1379 evidence supports the idea that Huddersfield was already more important in some respects than its neighbour, where only two men paid more than the usual 4d tax. By contrast Huddersfield had a wealthy merchant and several craftsmen; a smith, a wright, a shoemaker and a tailor. It had also taken into its boundaries the 'town' of Bradley, a former grange of Fountains Abbey. No doubt Huddersfield owed its local status to the fact that it was the parochial centre for several neighbouring townships, but it also had a favourable location at the junction of two major feeders of the river Colne and was on an important north-south highway.

Even so, we need to remember that both Almondbury and Huddersfield were

The intriguing mound at Hill House, not far from Bay Hall. Difficult to photograph now because of the proximity of buildings and trees, this image is from c. 1967, looking north. The facade of the 'Teapot' Chapel at the junction of Halifax Old Road and King Cliff Road is visible to the right. See Ark Hill Mound. Stephen Moorhouse collection

very small indeed at that time. In status and population the neighbouring town of Elland rated higher than either of them, and none of these can be compared in any way with more prestigious centres such as Wakefield and Pontefract.

The fact is that the tax roll of 1379 contains evidence of only 35 married couples in Huddersfield and 14 single people. However we use these statistics, and whatever allowance we make for children and people who may not have been taxed, it is clear that the town could have had only a very small population. Moreover, those few taxpayers would not all have been resident in the nucleated settlement, for we know from other sources that some of the families were living in Bradley, Deighton, Gledholt and Edgerton. It is likely also that the Brooks and possibly the Copleys were resident in the area across the town brook, not in the town itself. That area stretched away to Fixby, and would later be known as Fartown, although the parish boundary between the two may not have been firmly drawn in 1379. For example, surnames such as Hanson, Boothroyd and Lightridge, which had their origins and much of their history in Fixby, were all listed under Huddersfield, whereas Fixby itself was virtually depopulated.

National population figures do not allow us to presume that Huddersfield would have grown significantly through the 1400s and early 1500s. Only by the end of the 16th century, when we have information for almost forty significant place-names in the township, is there evidence to tell us exactly where many Huddersfield people were living. I have dealt with many of these sites in a number of other publications, but include their names here for the sake of completeness. On the other hand the town's street names have received little attention previously.

The Street Names

In A.H. Smith The Place-Names of the West Riding of Yorkshire (1961), the entries that have to do with Huddersfield contain numerous omissions and errors, and nowhere is that more apparent than in the paragraph entitled 'Street-Names'. Apart from a few 16th and 17th century references – none of them correctly identified - the evidence is late. Of the fourteen names listed, none has a correct reference earlier than the Market Place, noted in 1721. The only other date offered is for Folly Hall in 1843. Professor Smith may have been right when he said that Huddersfield has 'few street-names of any antiquity', but it has many that are of real significance and the topic deserves better treatment.

There are some old names. The town may originally have been no more than a small cluster of houses but these names offer a more intimate picture of the growing community by c.1600. At its heart were the church and parsonage, on opposite sides of a 'street' that was really part of the highway that linked Huddersfield to its neighbours. There was a well and several other buildings close to the church, although most of the land immediately to the north and south was under cultivation, including several major crofts and arable land that was still held in common. This tiny nucleus was set in an expansive 'green' where some enclosure was already taking place. In fact the green lay round the town on every side except the north, and each section had its own name. Several dwellings were sited on it, including Broad Tenter, Backside, Shore Head and Croft Head. Other named houses were Norbar, Fold, Hall and Smithy

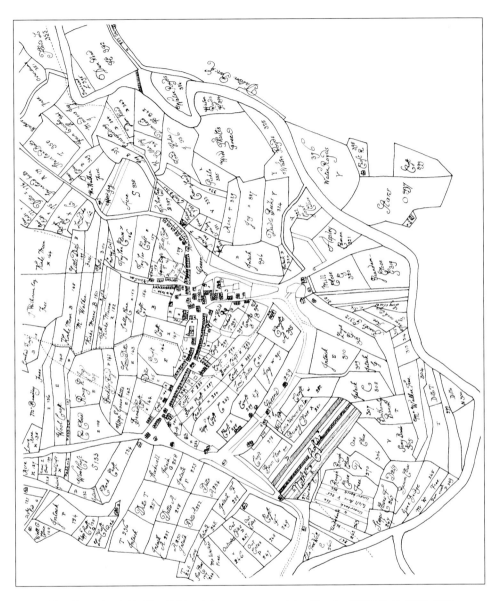

The Ramsdens' Huddersfield estate was surveyed by Timothy Oldfield in 1716. This small plan, copied from the original map by GR, shows the corn mill and its goit, the parish church, and the town street with its market place. No names are given, but Norbar Lane is at an angle to the street, to the east of the parish church, and the six dwellings of the Row are prominent at Hall Croft, alongside the town lane. The many 'Intacks' are land reclaimed from the town green, and the strips at Nether Crofts show that some agricultural land was still held in common.

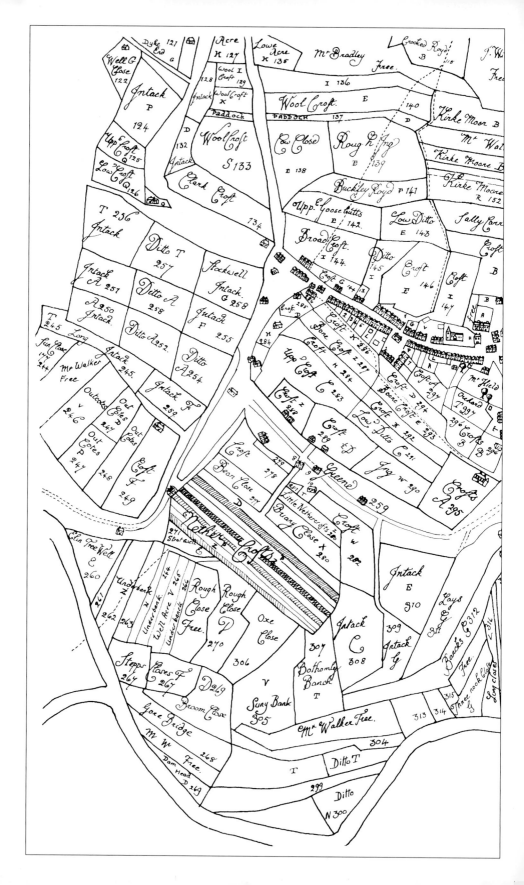

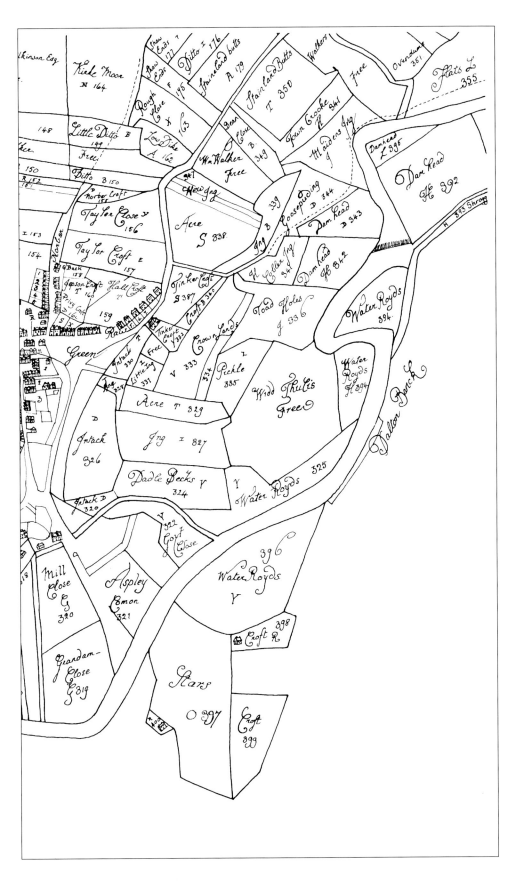

Hill, but these were probably located more centrally.

The area immediately to the east of the church, what we now call the Beast Market, seems to have been where most travellers arrived and departed. There was a lane which led down to the corn mill and the bridge over the river Colne, and this was also part of the highway to Wakefield, Pontefract and other more distant places. From the same location there was another lane that lead out to the town 'ings' or meadows. There it turned abruptly to the north-west, en route to places such as Halifax, Bradford and Leeds. Also running north, past Norbar, was a modest lane into the fields that would later assume greater importance. Close by was the manor house, known as Bay Hall.

The Ramsdens

It was about this time that a family called Ramsden came into possession of the manor of Huddersfield, a purchase that would see them play a major role in the town's development for more than three centuries. The story starts with William Ramsden, a yeoman who had made influential connections through his marriage, and acquired a significant estate in Almondbury and Huddersfield by 1542. He was able to build on that after the Dissolution, so that when he died, in 1580, the family was already in a powerful position locally. William's younger brother John significantly increased the family's holding, and his descendants were able to complete the purchase of both Huddersfield manor (1599) and Almondbury manor (1627).

Although the family left Almondbury towards the end of the 1600s, and went to live at Byram, near Knottingley, it was Huddersfield and Almondbury that continued to provide them with their main income and much of their energy went into developing the Huddersfield estate in particular. For example, in 1671, they secured market rights for the town, no doubt attracted by the revenue that such a move might generate. That was followed by the erection of the Cloth Hall in 1765, and the building of the canal from Huddersfield to Cooper Bridge, a project completed in 1780. It was in the latter part of this period that Huddersfield was linked to other growing towns and cities by improved turnpike roads: the enclosure of its commons took place in 1789, and the canal through to Ashton-under-Lyne in Lancashire was begun in 1794. All these developments had some influence on the town's place-names, and the maps and documents associated with them are of great importance.

Nevertheless, in the very last years of the 18th century, the town's existing names were still few in number and suitable only for a small community. The one and only thoroughfare, known colloquially as the town gate, continued to be Huddersfield's main axis, running west to east past the parish church. Strictly speaking this was not a place-name but simply the local word for the highway itself. Those who were keeping the town's records would have been aware of that and, from as early as 1657, we find it referred to more fashionably as the 'town street'. This appears to be the first use in Huddersfield of that word. In addition to this 'street' Huddersfield also had its Cloth Hall, a market place, and areas set aside for the trade in corn and cattle. Despite that, and despite the increased activity by the canal at Aspley, much had not changed in the town since 1600.

It still had a very small population and few public buildings of note. Many visitors were not impressed favourably by what they saw. In 1777 an American visitor described the town as 'very old built', with 'a wretched appearance'. Surprising as it may now sound, this small but vibrant residential and commercial centre, with its uncomplicated nomenclature, was still surrounded by fields and areas of waste as late as c.1800. The names of these open areas, both cultivated and uncultivated, would disappear in the next few decades, as the land was progressively covered with buildings; many new names would be coined, from several different sources, and the speed of change would mean that some had only a short life.

In fact, the experience that most local people had of name-giving was probably confined to their family, their animals and possibly their fields. A few of them would be familiar with the street names of Leeds and York, and others would have some knowledge of London, but they probably had few ideas about how the new streets and buildings should be named. They were, in any case, tenants of the Ramsden family who owned most of the town and who clearly thought that name-giving was their prerogative. That attitude is implicit in the earliest choices that were made, and actually explicit in estate records and the Improvement Act of 1848:

And be it enacted, That it shall be lawful for the Lord of the Manor to determine the Name or Names by which any new Street or Streets to be hereafter formed upon the Property and Estates of the said Lord, or upon any Estates hereafter to be purchased and annexed thereto, shall be called or known.

In one sense the naming of the streets was a luxury, for the more pressing need was to bring the town into the 19th century. The task in front of the community is expressed in the earlier Act of 1820 which was concerned with the lighting, watching and cleansing of the streets. Commissioners had been appointed, and they included, as might be expected, several members of the Ramsden family, together with innkeepers, bankers, tradesmen and manufacturers, all of them men with an annual personal estate of at least £1,000. The limited powers of the commissioners allowed them to authorise gas works, have gas pipes laid, and lamps set up in the streets and alleys. The water works already belonged to the Ramsdens, so the concern there was to protect them from ill use or vandalism. Able-bodied men were to be employed as watchmen and given powers of arrest, the streets would be watered and carts provided for the removal of litter, ashes, dirt and horse dung. Those employed to do that work, scavengers as they were initially called, would ring a bell or give a shout to warn of their approach, so that people living in inaccessible alleys or yards would have time to bring out their rubbish. The Act contains much more on all these matters, and on how money would be raised to pay for the improvements, but indirectly it tells us also how rapidly the urban area had grown in just a quarter of a century.

The early 19th century

The estate maps that were drawn up between 1778 and 1797 demonstrate the significant changes that were beginning to take place in Huddersfield and they are followed by a number of town maps produced in the period 1818-26. These differ in some details, but basically they all tell the same story. In George Crosland's map, of 1826, the importance of the town gate is still obvious but it has been renamed Kirk Gate and its extension westwards has become West Gate. Norbar has been transformed into Northgate and, at the bottom end of the town, are Castle Gate and Water Gate. Huddersfield had suddenly acquired a central corpus of names that brought it into line with its neighbours, Leeds, Bradford, Pontefract and Wakefield. The difference is that the 'gates' in those towns and cities were all ancient names.

In 1878 John Hanson made the significant comment that Huddersfield's streets were named at the turn of the century, that is c.1800, and the evidence bears out what he said. Nevertheless, the way in which some of the major names changed, from one estate map to another, suggests that they were the subject of much discussion, and a few of those first given 'street' as a suffix were scrapped. There had clearly been some hesitation as to whether the northern word 'gate' should be employed and the decision to use it suggests that the Ramsdens had decided to give the town a core of historic names, allowing it to be compared with its neighbours.

However, by 1826 the new term street was already numerically more important than gate, and it had been widely used in the developments to the south of the old town gate. Cloth Hall Street and King Street now provided an alternative link between the top and bottom of the town, and parts of the route across the Back Green had become High Street and Ramsden Street. Running north and south, and either crossing these streets or linking them, were Upperhead Row, Market Street, New Street and Queen Street, forming a rough grid pattern that was criss-crossed by other streets and alleys. The number of amenities and public buildings had also grown, with a new church and several chapels, the riding school, baths, a gas house and a lock-up. Aspley, between the river and the canal, had become a much busier place, with a timber yard, a foundry, warehouses, a toll bar, mills and the *Wharf Inn*.

In the town centre some fields and open spaces remained, a few still under cultivation and others serving as tenter grounds, bowling greens and rope works. There were even market gardens where St Andrews Road now is, and the town pinfold was sited near Union Street, an evocative link with the past. Nevertheless, there is little evidence on the map of intensive development to the north of the former town gate, and the *George Inn*, which was the town's premier hostelry, still occupied the north side of the Market Place, blocking any continuation of New Street to the north. Of real interest though is a rough grid of 'intended roads', between the New House estate and the turnpike to Leeds, across what was later St John's Road, and with North Gate running through the centre of it. The words New Town are written boldly and stylishly across what is now Cambridge Road and Clare Hill. The map captures Huddersfield at a fascinating moment in its history.

The major streets of the early 1800s were named by the Ramsdens and

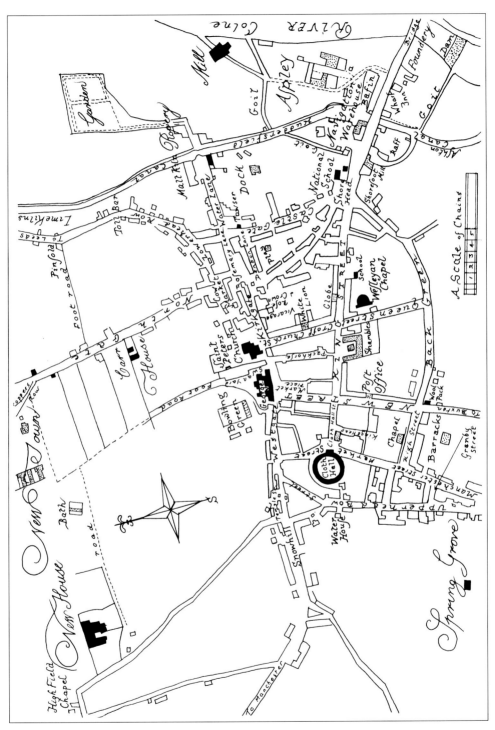

A Plan of Huddersfield in 1820, redrawn by GR from a lithograph of 1903. On this early and rather crude map are the new street names, inns, public buildings and industrial sites. Development has taken place at the Cloth Hall and canal side, but foot roads, the Back Green and the pinfold recall the old town.

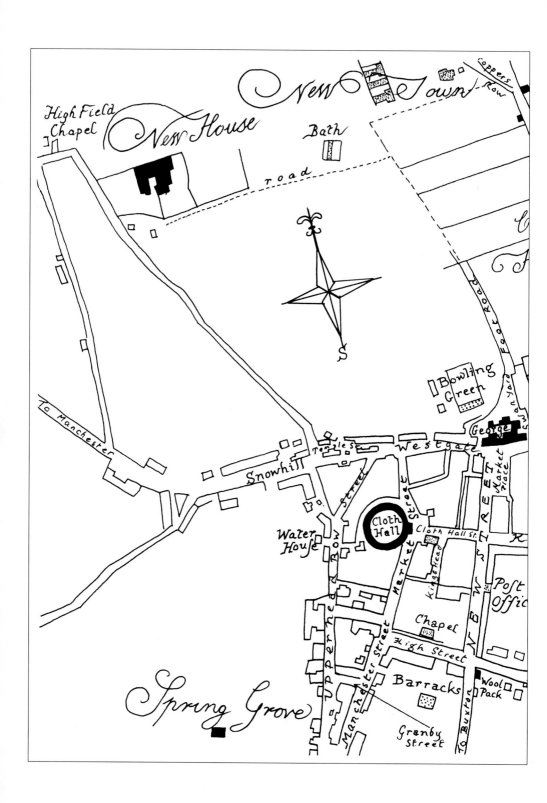

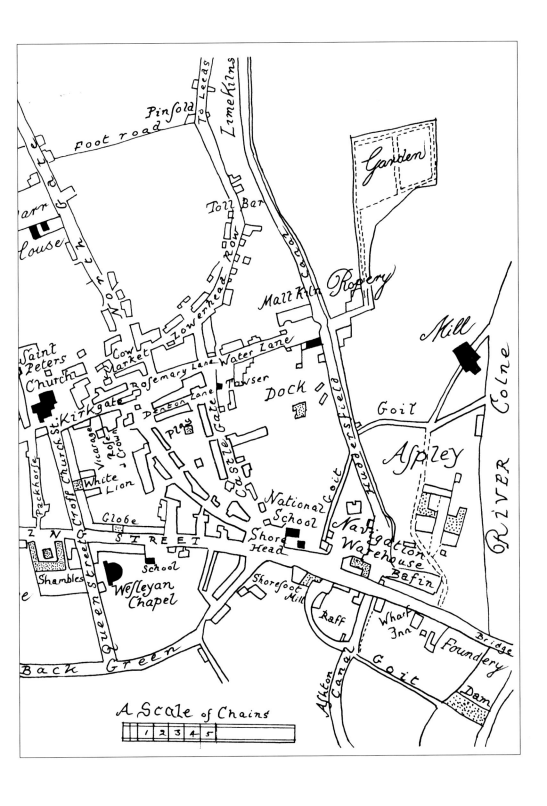
A Scale of Chains

those who acted for them, and they reflect the family's power and aspirations. Many of them have survived and yet the town centre names as a whole contain evidence of other major influences.

Inns and public houses
The first influence is that of the town's numerous inns and public houses, places referred to frequently from as early as the 1600s. Unfortunately the names and signs are not given until much later. An entry in the Quarter Sessions for 1675 is typical in that it refers simply to events 'in Gervase Kaye's house'. In such records 'house' usually stands for public house, as when a clothier told the court in 1729 that he had returned to the market leaving his 'piece of plain cloth ... in a Closet in the house of Joseph Sykes, innkeeper'. It is particularly frustrating not to know the name in circumstances where the location of the house would be of real interest. For example, Susan Berry was found guilty of a felony in 1729 and the court ordered that she 'be whipt at Huddersfield ... the next Market Day, betwixt the houres of twelve and one ... from Michael Bramhall's house ... down to the Mills'. This is likely to have been the *Queen's Head*, on the west side of the Market Place, referred to by that name just a few years later when a man called Bramhall was the tenant.

Other early names, such as the *George* (1713) and the *Swan* (1754), also belonged to a long national tradition and seem unlikely to have been influenced by the Ramsdens. Much the same can be said of the licensed houses listed at the Brewster Sessions in 1777: *Blackamoor, Rose and Crown, Maysons Arms, Bulls Head, Clothiers, Tawbut* (Talbot), *Hors shoos, Fleece, Horn, Packhors, White Swan, White Bear, Brown Cow, George, Queens Head, Cross Keys, Blue Ball, Saddle, Plow, White Hart, Redd Lion, Boot and Shoe, Hatters, Shoulder of Mutton*. There were two *White Lions*.

In the Ramsden estate survey of the following year, there are several names not in the above list. These were the *Spread Eagle*, the *Blacksmiths Arms* at the bottom of town, the *Hat*, possibly identical with the *Hatters* above, and two inns in the Market Place, the *Horse and Jockey* and the *Kings Arms*. The map that accompanies the survey shows us exactly where these were located, with the biggest concentrations in the market areas. They were all traditional names and I find no reference to the *Ramsdens Arms* before 1795. Some of the beer houses or 'Tom and Jerrys' may have had more local names, for example the *Wasps' Nest* and *Noah's Ark*.

The Yards, 1837
A second influence is evident in the street directory of 1837 which shows that building development off the streets, in yards and folds, was giving rise to a body of quite different names. A few of those buildings have managed to survive more recent changes in the town centre and whereas one or two sound like nicknames most are a direct link with families who made their mark in the town in the 1800s, for example Battye's Buildings, Hawksby's Court and Hammond's Yard. These places were not consciously named in the way that the streets were, but started out as simple associations with important tenants, builders or owners. Inevitably, many had only a short life but some survived, and they

represent a phase in Huddersfield's street history that is worth recording. Even if we omit the names of the yards associated with particular inns, the list in the directory for 1837 is still impressive:

Yards: Barker's Yard, Battye's Yard, Bottomley's Yard, Boulder Yard, Brook's Yard, Eastwood's Yard, Gibson's Yard, Helliwell's Yard, Horsfall's Yard, Howard's Yard, Kilner's Yard, Lockwood's Yard, Lumb's Yard, Marshalls Yards, North's Yard, Rhodes' Yard, Roberts Yard, Roebuck's Yard, School Yard, Skilbeck's Yard, Sykes' Yard, Thornton's Yard, Waterhouse Yard.
Buildings: Bath Buildings, Bradley's Buildings, Brook's Buildings, Carver's Buildings, Helm's Buildings, Johnson's Buildings, Jowitt's Buildings, Lancaster's Buildings, Schofield's Buildings.
Courts: Booth's Court, Hamer's Court, Hampton's Court, Windsor Court.
Squares: Commercial Square, Fenton Square, Milton Square.
Folds: Chadwick's Fold, Walker's Fold.

The history of these generic terms, if not the etymology, is of real interest, and a few of the names will be looked at in the dictionary section. However, it is quite clear that 'yard' was the most frequent of the new generics, so we should, I believe, give some thought to its predominance. For example 'fold' was already in use in some neighbouring villages, notably in Honley but also in places such as Lepton where 'yard' was unknown. It may simply be that 'yard' owed its popularity in the town centre to the precedent offered by the names given to the inn yards. These partially-enclosed spaces were a familiar feature of the town, even before 1800, and they were not always adjacent to the inns they served. The Swan Yard, for example, was on the opposite side of the street to the *Swan Inn*, roughly along the line of today's Byram Street. Even as early as 1778 it was not given over entirely to stabling: there were several private houses, Martha Clayton's bakery, William Wilson's barber's shop, and the smithy belonging to Thomas Sinkinson. Furthermore, it was so closely associated with the landlord of the inn that it was sometimes called Dransfield's Yard.

The much less common 'court' is the word that various Acts of Parliament used, along with 'street', for the new developments in the town, so it is hardly surprising to find it here. However, it may have sounded rather more impressive than 'yard', raising the status of the buildings. I am inclined to that view because Hampton's Court and Windsor Court seem not to be surnames, although that would be possible, but wishfully or ironically associated with the Royal family. King Street and Queen Street were near by. 'Buildings' was a logical and fairly predictable term, but 'square' was certainly a deliberate borrowing, designed to give a group of houses a touch of class. It did not appear to matter if the houses were not actually in a square. Later, many more terms would be employed for which there were no local precedents, e.g. arcade, avenue, close, crescent, drive, parade, place and terrace, and the timing and popularity of all these choices are worth noting.

However, where the town centre is concerned it is the names of the yards and courts that first offer a new perspective, for they reflect Huddersfield's development at a different social level. The sometimes temporary nature of these names and the circumstances in which they developed were mentioned earlier, and there is certainly no evidence that the Ramsden family or their agents had anything to do with their choice. It might be thought that they attached more importance to the appearance of the town, and the rent book, than the condition of the dwellings hidden away behind the splendid new façades.

The Villas

If the consciously-named streets reflect the aspirations and self-importance of the Ramsden family, and the unstable names of the yards identify the ordinary townsmen, then the new names given to villas and 'cottages' are revealing about the status that wealthy and successful citizens attached to name giving.

From c.1790 large and imposing houses were being built very close to the town centre, notably Mr Fenton's Spring Grove and Mr Horsfall's Thornton Lodge. The generics 'grove' and 'lodge' were to be very influential locally and would be used in scores of 19th-century villas in Edgerton, Birkby and the New North Road area, almost obligatory choices for successful manufacturers, tradesmen, bankers and solicitors. They were not the only villa names but they were more distinctive than 'house' and 'cottage' which also became quite popular.

As the town continued to grow many of these prestige houses found themselves under pressure from the expansion and were either demolished or suffered a loss of status. Their names though often remained in use, identifying roads and streets or even whole districts. For example, few local people now associate the name Thornton Lodge with Mr Horsfall's house, even though it has survived in an area of high-density housing. As far as I am aware the importance of this category of names has not previously been commented on.

PLACES

ALBERT HOTEL, ALBERT MILLS, ALBERT STREET, ALBERT YARD

We tend to think of Albert as a traditional English first name, now out of fashion, and yet it came into use only after Prince Albert married Queen Victoria. In fact, Victoria had also been an uncommon name but, after the royal marriage, both became popular choices, not only for children but also for inns, mills and streets. In Huddersfield there was an *Albert Inn* in the Albert Buildings on New Street in 1853, and then, in 1878-79, the architect Edward Hughes designed the present *Albert Hotel*. Appropriately it is on Victoria Lane which had Victoria Buildings to the west, and both these names are on the town map of 1852. Lockwood also has its Albert Street, Albert Mills and Victoria Street.

ALBION INN, ALBION STREET

This street was part of the early 19th century development that followed on the creation of New Street and Buxton Road, and the name is listed in 1822. It is on the town map of 1826 and is almost certainly linked with the *Albion Inn*. Francis Dalton was the landlord there in 1837. At that time there was also a public house called the *Albion* at Longroyd Bridge. A local diarist records how a man was killed in Albion Street in 1850 when a 'skip' fell from a crane as he was 'passing on the footpath'. Albion is a poetic name for Britain, and it derives from the Latin word for 'white'; it is thought to be a reference to the chalk cliffs of England's south coast – the invaders' first view of the country. The white cliffs feature on some inn signs, but ships were also named Albion and that explains why some inns have signs showing a fully-rigged sailing vessel. An early example of the inn name is the *Albion Hotel* in Chester, founded in 1715.

ALDER STREET

Alder Street is much longer now than it was formerly, and the extension to the north may have come about after the farm at Flash House was demolished. The name and original location seem to link it with the adjoining terraces which were also named after trees. At the junction of Alder Street and Calton Street is a triangle of open ground: it was the subject in 1880 of a minor disagreement between the Commissioners and Major Graham, the Ramsdens' local agent. He had offered them the plot 'as a permanent open space' but there were conditions that related to the adjoining streets and the Commissioners declined the offer. They informed the Major that he was at liberty to use the land as he deemed proper. It seems fitting that the Council has now landscaped the site. The short row of houses to the north is Hebble Street.

ALFRED STREET

Sir John Ramsden's agent wrote to the Commissioners in October, 1865, saying that it was their intention 'to lay out the new street called Alfred Street'. I feel sure that Alfred would have been the family name of a kinsman or close acquaintance but I have been unable to identify the connection. The directories suggest that it was never intended as a residential street and in 1879 almost all the premises had commercial or industrial occupiers.

ALLISON DRIVE, ALLISON DYKE (Fartown)

This street name has associations that take its history back some 300 years. In 1671, Anthony Allison of Huddersfield sent a petition to the Quarter Sessions in which he described himself as a poor man with a large family who had 'sustained diverse losses, and was likely to bee put out of his house'. He was seeking permission 'to build a cottage upon the common ... called Naithroyd Hill', and the request was granted. Its exact location is not known but when a local woman was accused in 1679 of stealing ducks in Huddersfield, she described how she had walked out of the town to Anthony Allison's house to ask him to dress and roast the birds for her. At that time his cottage would have been relatively isolated, probably close to Allison Dyke at Netheroyd Hill. This stream is first mentioned in the surveyors' accounts of 1788 and later shown on OS maps.

AMEN CORNER (Kirkgate)

This was probably an 'unofficial' name and it is not well documented although a picture of it has survived. It apparently referred to an area just east of the church, where there was formerly a gate into the churchyard and where a man called Hart had a chemist's shop. Just across the street was the vicarage. Edward Law notes that Isaac Hordern described the old *White Horse Inn* as being at Amen Corner in 1854.

ARK HILL MOUND (Birkby)

The mound referred to here is located close to Beacon Street and it is probably the 'Mount' shown on the OS map of 1843. Archaeologists have suggested that it may have been a Norman motte. Philip Ahier discussed four or five possible explanations of the name in his Legends and Traditions of Huddersfield (1941-42) but conceded that much of what he was repeating was pure speculation. One theory was that the hill took its name from a nearby public house called *Noah's Ark*, kept by Aquila Priestley. What Ahier did not know at that time was that this property had been the subject of a photograph in 1863, part of the evidence offered in the town's long-drawn out tenant-right case. It was described there as 'a building called *Noah's Ark*, Hillhouse, built by Aquila Priestley fifty years ago' - that is in 1813. It had a single storey and was in very poor condition. I have found no evidence that Noah's Ark was a public house, and there is nothing in the photograph to suggest that it was, but it was certainly a beer house and it is listed in at least one trade directory as *Noah's Ark*.

ASH BROW (Fartown)

Solitary trees were often used in the past as boundary markers or to identify settlement sites. One early site is *Grange Ash* near Denby Grange, shown on Saxton's map of Yorkshire in 1577. A typical boundary marker was the 'ashe tree ... felled and cut down within the memory of man' in 1625; it was at the heart of a dispute over pasture rights in Rastrick. The place-name Ash Brow probably goes back to the 16th century, when the Mellor family of Cuckold's Clough possessed land 'latlye enclosed of the common'. In a deed of 1633 this enclosure was described as 'lying upon a place called Sheepridge, of which ten acres doth abut upon Meller Eshe', that is the Mellors' ash tree. By 1789, when the common was finally enclosed by Act of Parliament, this area was known as 'the common called Ashbrow', and today this is the name of the inn which stands by the side of Bradford Road. It appears to have contributed to the decline of the much older place-name Cuckold's Clough.

ASPLEY

This was originally the low-lying area by the river Colne, close to what we now call Somerset Bridge, but on the Dalton side of the river. The inference is that the name originally referred to a clearance where there were asps or aspens, and this would certainly suit the location. It has a long history. In 1454, for example, William Blakbourne of Dalton was indicted at the manor court for not cutting back his hedges in 'Asppelay Loyne', whilst a title deed for Storth, dated 1566, mentions 'Asplebrooke'. However, the suffix 'ley' suggests that the place-name may be much older, possibly going back to the pre-Conquest period. The river Colne changed course in that part of Huddersfield over the centuries, and this may in part explain how Aspley ceased to be thought of as a Dalton place-name. Town maps show that the land there was largely undeveloped and Aspley Common was the name given to an area between the tail goit of Shorefoot Mill and the river. The canal brought changes to that part of the town after 1775, and later maps show a mill, warehouses, a bowling green and substantial houses on the site. It became a desirable place to live and several villas were built there in the early 1800s. In 1837 these included Aspley

The Somerset Arms, *Aspley; a name influenced by the re-named Somerset Bridge.*
G Redmonds

House, the home of Mrs Mary Ann Atkinson, and Aspley Place, where Sir John Ramsden's agent lived. By 1849 it was a place of industry, despite the presence of Aspley Cottage, and there were coal yards, weighing machines, print works, a brewery and the *Wharf Inn*. This public house is on an earlier map of 1818. Nevertheless, Lower Aspley House was still a villa residence, with statues in the formal gardens.

BACK GREEN
Before Huddersfield's rapid growth in the 19th century, when most of the houses lay along what we now call Kirkgate, there were cultivated fields immediately to the west of that street called the Bone Crofts. Beyond them was an area of waste or common known as Backside and that name gave way to Back Green in the 1700s. Back Green is shown on the estate map of 1778 and it remained in use well into the 1800s. In the Huddersfield Enclosure award of 1789, the Wakefield to Austerlands turnpike, the road to Manchester that is, was described as running from Huddersfield Bridge to Outcote Bank, across the Back Green, a highway 'of breadth 40 ft between the ditches'. This was roughly on the line of Ramsden Street and High Street and their development was part of the reason for the decline of Back Green as a place-name. Nevertheless, it was still being used during the Oastler era, when it was the scene of large open-air meetings and torchlight processions. I heard the name used by one old lady in the 1970s.

BACKSIDE, BACK OF THE TOWN
This was the original name of uncultivated land in the area now occupied by the Town Hall and the adjoining streets. A document of 1571 refers to 'the lately inclosed common lands of Huddersfield called Overgreen, Neythergreen and Backside'. However, it had several values as a place-name and is first recorded in 1548 as 'a capital messuage'. This had been purchased by a merchant called Richard Charlesworth from James Boothroyd of Rastrick, and it was probably given the name because it was the first house to be built on the waste there, literally on the 'back side' of the town. By 1600 other houses had been built and several families were said to be 'of Backside'. The name now had territorial connotations, referring to a few scattered dwellings: as their number increased the inhabitants came to be thought of as a community and versions of the place-name remained in use well into the 1700s. In 1716, for example, it was 'the Back of the Town' and in a trade directory of 1760-63 a schoolmaster called Joseph Collingwood was said to be of 'Back side town'.

BANKFIELD ROAD (Outcote Bank)
There were some thirty names entered for Bankfield Road in the street directory of 1879 and the houses were of a good quality. Several had been given names by their occupants, for example Bank Villa and Kirk View. The development on this site took place after 1857, when the trustees of the Fenton estate thought of 'opening out a centre road from the bottom of Outcote Bank'. Jonathan Brook was the tenant at that time and the original plan was to drive the road through his fields 'along the north side of the church'. The trustees were conscious

of the potentially 'detrimental effect of smoke from Longroyd Bridge' but the development still went ahead. From midway along the road there is a precipitous flight of steps up the bank to Spring Grove, where the Fentons formerly lived.

BANK TOP

This was the top of Outcote Bank. John Hirst had a new cottage at Top o'th' Bank in 1768-71, and the place-name was still in use in street directories in 1822. It is shown on Crosland's map of 1826.

BARRS LANE (Lost)

This name has a long if obscure history. The lane is marked on the canal plan of 1773, running east from the new turnpike road to Leeds, roughly in the area where Gas Works Street now is. Over a century earlier, in 1664, George Batley was accused of not cleansing his ditches between 'Feildyate and Barrslane Head', a description that puts it in much the same position. It may have been the lane shown on the map of 1716, running from Tinker Croft through a field called Goose Pudding, and then becoming a footway along the riverside. With 'Barrs' as the prefix it may have had some connection with Norbar.

BATH BUILDINGS, BATH STREET

Little is known about the baths which were responsible for these names although they are marked on town maps of 1820 and 1826 and were presumably public baths. There can be no connection though with the Cambridge Road Baths, opened in 1930. Bath Buildings was a place-name from before 1828, and Bath Buildings Lane from the early 1860s. In today's Bath Street there is an impressive building with a fascinating history. It was opened as the Hall of Science in 1839, by the followers of Robert Owen, and was intended as a centre for education. Sadly, for the Socialists that is, the enterprise soon failed and the premises were used as a chapel, first by Unitarians and then by Baptists. Now, the building bears the sign of the painting contractor Ramsay Clay and the street is virtually a backwater, unfamiliar to most Huddersfield people.

BATTYE'S YARD (Market Place)

Little has survived of this yard which was on the east side of the Market Place next to Barclay's Bank. The site is now marked by a short alley, but a fine drawing by Noel Spencer reminds us of how it looked until quite recently. It is said to have taken its name from John Battye, an attorney who went to live there in 1805. His address is given as New Street in 1809 but he appears to have moved soon afterwards to Queen Street. The place-name itself is in numerous directories and in 1837 it was listed in the Market Place. It was the location of half a dozen small businesses and the Amateur Artistes Society had its headquarters there in 1879.

BAY HALL (Birkby)

Much has been written about Bay Hall, partly because the small, timber-framed house is a rare survival from the 16th century, and partly because the description of it as a 'hall' hints at an important past. And yet all we know is that a family called Brook was living there in the 1560s, as tenants of the Byrons who held half the manor. When Edmund Brooke made his will, in 1573, he referred to several properties that he had 'late boughte of John Byron, Esquire' and to his 'Capitall messuage caulled Bayhall', a property that he still held on a lease. One of the attractive features of his will is the number of distinctive Huddersfield field-names that he refers to. However, the Byrons were having money problems by 1573 and they had to relinquish ownership of the hall soon afterwards. Because the Brooks were tanners it is tempting to see a connection here with 'John Brooke of the Barkhouse' who was taxed 12d in 1524. A bark house was a place where tanners stored the bark they needed, so perhaps this was an indirect reference to the same property. The interpretation of this place-name is unfortunately not clear. 'Bay' is a word which has several quite different meanings and there is nothing in the earliest evidence to indicate which one it might be.

BEACON STREET (Hill House)
This street was built after the demolition of the old farm known as Hill House, probably in the early 1870s. There is no evidence to suggest that there was ever a beacon on the site or close to it but the conical shape of the adjoining mound or motte may have suggested the idea. David Shore has suggested that the line of the street may follow the boundary of a bailey. See Ark Hill Mound.

BEAST MARKET
When Huddersfield had its right to a weekly market granted in 1671, one or two areas close to the town's main street were set aside for trading, possibly immediately. There is no conclusive evidence that today's Beast Market marks the original site where cattle were bought and sold, but it seems likely that it was. The name is written on a map of 1778, although probably not in a contemporary hand, and it is clearly shown as an open space in c.1780, at the very bottom of the area we still know as the Beast Market. On that map, the lane that leads towards the church is named Cowgate, a lane so narrow, according to D.F.E. Sykes, 'that two carts could not pass'. In fact Cow Market was a common alternative to Beast Market in documents of the early 1800s but this may be a scribal preference only: 'beast' is the usual dialect word locally for cattle. The streets there would still have echoed to the noise of cattle in the 19th century and it was decided to erect railings on the pavements in 1850 'to protect foot passengers … on the Market and Fair days'.

BELGRAVE TERRACE
I cannot date these handsome houses exactly, but the terrace must have been built in the late 1830s. It is not mentioned in the directory of 1837 but the Lighting Commissioners called for a lamp to be placed there in 1839, and it was there when the College was being built in 1838-40. The writer of an article in the College Magazine for 1878 recalled that 'there were no houses from the Catholic Chapel (except York House) until the terrace of buildings…
…which extends to the College gate'. The New North Road area was then a very fashionable part of the town and that is reflected in the name Belgrave.

This may have been inspired by London's Belgrave Square, which in turn was named by the Grosvenor family, as Dukes of Westminster. Belgrave in Cheshire was another of their properties. Belgrave Villa was the home of Joseph Bates in 1879.

BELMONT STREET
Belmont, like Beaumont, is of French origin and means 'beautiful hill'. There are many Belmonts throughout the British Isles, but most of the places so called are gentry houses, like Belmont House in Doncaster and Belmont Hall in Cheshire. The likelihood is, therefore, that the word had the right social cachet as villas and terraces sprang up in the 19th century. The council minutes of August 1881 record a resolution 'that the name of Brunswick Place be changed to Belmont Street, and that the concurrence of Major Graham in this alteration be obtained'. Major Graham was acting for the landlord, John William Ramsden, who had to be consulted when a new name was given but the use here of 'obtained' rather than 'sought' suggests how the relationship operated. Close by were several detached houses with gardens, called Greenfield, Bent's House and Bent's Cottage. They are shown on the first OS maps, and a gentleman called Thomas Ibbetson was living at Bent's House in 1837.

BIRKBY
On the face of it Birkby should be of Scandinavian origin, with 'birk' as the local form of birch and the suffix 'by' pointing to a pre-Conquest settlement. However, the place-name is unusually difficult, partly because the evidence is relatively late and partly because there are several contradictory spellings. These are likely to be scribal errors but they help to emphasise the name's early obscurity. In 1555, for example, Robert Clay was said to be 'of Kirkebye' in Huddersfield, whilst on the estate map of 1716 the name is written as 'Birkwith'. A trade directory of 1879 has Birkley, and Birtby is found in the Elland parish register in 1696. The earliest spelling that I have found is in a faded remnant of court roll, dating to 1532, and there too it may be written Kirkby. Certainly the name had two values, referring initially to a single house and then to a wider territory as the number of dwellings in that part of the township increased. No more than four or five families were resident there in the 16th and 17th centuries.

BIRKBY HALL ROAD
What we now call Birkby Hall Road was previously Birkby Lane and it was said in a Quarter Sessions indictment of 1771 to lead from 'Hillhouse to Birkby Lane Head', part of the highway to Elland. It is not clear what we should infer from its description as a hall. There is a half-timbered house on the north side of Birkby Hall Road, substantial enough perhaps to have been a yeoman's 'hall'. However, on the OS map of 1854 the name Birkby Hall seems to refer to a house on the south side, where Birkby Grange appears on later maps. In 1882, when the road was widened, the improvements began near 'the entrance gates to Birkby Hall'.

Birkby Hall Road in the early 1900s: the white-painted house still stands and has a timber frame   Clifford Stephenson collection

Birkby Hall Road, during the demolition of the old house next to Joe Fletcher's: he was a nurseryman and gardener. Compare the photograph on p. 26   G Redmonds collection

BIRKBY LODGE ROAD

Like Birkby Hall Road this name is found on the OS map of 1854 and it almost certainly derives from the name of a villa, probably the home of Thomas Holroyd. When Thomas died, in 1833, he made provision in his will for the education of five poor children of Fartown and founded the almshouses in Birkby Fold. He and his parents are commemorated on a tablet in the parish church. What we now call Birkby Lodge Road was simply Birkby Road in the Enclosure Award of 1789, 'leading westwards to the east end of Birkby Lane'. In 1882 it was said to include 'the bridge ... over the stream called Clayton Dike'.

BIRKHOUSE LANE (Paddock)

'Birk' was the usual term locally for the birch tree and it features in numerous place-names. Indeed, there were several early settlements called Birkhouse, enough to make them difficult to identify just by the place-name. Fortunately, the farm that gave its name to this Birkhouse Lane is sometimes identified for us in the parish register where it is said to lie near Longroyd Bridge. It was actually on the Lockwood side of the river and Thomas Sykes was living there in 1667. He was followed by Richard Hanson in 1700. There was a Birkhouse Terrace on Manchester Road in the 1860s.

BLACKAMOOR

There was an inn known as the *Blackamoor* in the Corn Market and it is listed as a Huddersfield public house in the Brewster Sessions of 1777. The landlord was called James Booth, one of several local publicans with that surname. In the 17th and 18th centuries it was not uncommon for rich families to have a Negro or 'blackamoor' as a servant, and it became a popular tavern name. Locally, Dame Savile of Bradley in Stainland had a blackamoor serving girl called Isabella 'Aethiops' in 1601; Daniel Whitley, also an 'Ethiopian', was a servant of Mr Beaumont of Whitley in 1782. The *Blackamoor* may have given way to the *Boy and Barrel* by c.1800.

BLACK DYKE (Fartown)

This farm was located on Fartown Green, in the north part of the angle formed by the junction of the roads to Bradford and Sheepridge. The name means 'black stream', but otherwise we know little about the farm, except that a branch of the Brook family held the tenancy in the 17th century. There are, for example, several references to a John Brooke 'of Blackdike', starting with the hearth tax returns of 1664. In 1681, three cows were 'rescued' from the pinfold in Huddersfield by a man with that name.

BLACKER ROAD (Edgerton)

Blacker Lane, as it was formerly called, was once part of the highway which ran from Longroyd Bridge to Jack Bridge in Birkby. It was almost certainly an ancient route but the present name goes back only to the period 1580-1680, when a family called Blacker lived in that part of Huddersfield. In 1631, for example, they were tenants of a house at 'Blackeryate', that is Blacker Gate. This is not a reference to the highway but to a gate across the lane which may have marked the point where the family's land butted up against Marsh Common. Their fields were referred to as Blacker Crofts in 1677. In 1878 an

The *Spink* Nest Inn, *Blacker Road: the name survives but this older house was demolished in the 1930s.*   Clifford Stephenson collection

old boy of Huddersfield College remembered the lane from his schooldays in the 1840s, and lamented the changes that had taken place: 'at one side of the field was the lane, now sadly shorn of its homely beauty, endeared to us by the name of Blagger Lane'. In 1882 it was being called 'Blacker Lane otherwise Blacker Road'.

BLACK HOUSE (Fartown)
'Black' was a common element in minor place-names, particularly those referring to water, but it seems unusual in the name of a house. It may be, therefore, that the earliest connection in this case is with nearby Blackhouse Brook or Dike: the farm Black Dyke was only a few hundred yards away. Edmond Brooke was living at the 'Blakhouse' in Fartown in 1524, one of several branches of a family that was exceptionally numerous in Huddersfield. In fact, of twenty-six individuals taxed in the town that year, no fewer than ten bore the surname Brook. See Canker Lane.

BLACK SHAW (Lost)
This name is something of a mystery. It is on Jeffery's map of 1772 as Blake Shaw, in Marsh, somewhere near the bottom end of today's Eldon Road. It is recorded in the court rolls as early as 1658. There is also 'Blakshaw' on George Crosland's map of 1826 although he gave the name to a locality near Dyke End.

BONE CROFT(S) (King Street)
Until the end of the 18th century, the land to the south of today's Kirkgate was still not built on, and the fields there had the name Bone Croft or Boone Croft. In the Middle Ages 'boons' were services owed to the lord of the manor, so the inference may be that these town centre crofts were worked originally by the tenants, as a tenurial obligation. They almost certainly date to well before the 1500s when the name is first recorded. In 1523, for example, Richard Beaumont of Whitley Hall granted land to 'Richard Boderoyd', and the wording of the deed emphasises the rural nature of that part of the town. The 'parcel of land' being conveyed was said to be bounded by 'Bonecroft' on the east, the common on the west, a recent enclosure on the north and a close belonging to Richard Beaumont on the south. Much of that part of Huddersfield is now taken up by King Street and Cloth Hall Street.

BOOTHROYDS (Birkby)
The Boothroyd family has been prominent in Huddersfield from the 14th century and in 1578 'John Botherode' was granted a new lease of a messuage in 'Byrkebye'. In 1622-24, when Birkby still consisted of no more than half a dozen dwellings, this property was referred to simply as 'Botheroides'. It was not unusual to name a house after a tenant in that period but 'Boothroyds' did not stabilise and we cannot be certain exactly where it was.

BOTTOM OF TOWN
When Huddersfield had just a single important street, running downhill from

the west, the area below the parish church came to be known as 'the bottom of (the) town'. This term occurred regularly into the 1800s in rentals and the parish registers and variations of it are still used colloquially. In 1716 it included the houses down to Shore Head. As late as 1992, the local newspaper quoted an old lady from Dalton who was reminiscing about Huddersfield in her youth. One of the interesting things she said was that 'folk who lived at the bottom, down in Castlegate, were the bottom-enders'.

BOULDER YARD (Kirkgate)
This yard was on the east side of lower Kirkgate and it is listed in the directory of 1837. It is one of the earliest private yards to be mentioned by the Lighting Commissioners, who required a lamp to be placed there in 1838. We should picture it as a workaday place, since the residents in 1879 were a rag merchant, a cooper and a shoemaker. The word 'boulder' was used at that time for stones rather larger than pebbles but nowhere near as big as the boulders referred to by geologists, so the inference is that the yard was paved. In the earliest part of the century that was probably unusual enough to justify the name.

BOW STREET
This is now a comparatively short street that leads from Springwood Avenue to Spring Grove School, whereas it originally followed the line of the drive from Outcote Bank to the mansion at Spring Grove. The name described the curve of the drive round the headland, an origin it has in common with London's Bow Street. There is an ancient right of way from the top of the town to Paddock Foot, known as the Springwood footpath, and this starts now in Bow Street. When the tunnel through to Spring Wood was built it interrupted this footpath and the Commissioners were in dispute with the railway company about it for years. In 1848, for example, the company was required to have the footpath through the Spring Wood made passable and to put up and maintain lights near to the rubbish and the holes they had made. They argued in particular about the footbridge over the railway and this was not completed until 1858.

BOY AND BARREL
This well-known public house, with its very distinctive name and sign, has a long history in the town. It is listed in the Quarter Sessions records from 1803 and in street directories from 1822. It was an inn frequented by clothiers and has survived the modern urge to re-name old inns. Bacchus was the Roman god of wine, and on tavern signs he was depicted as a fat little boy sitting on a barrel. There are other public houses with the name *Bacchus* and Selby also has a *Boy and Barrel*.

BRACKEN HALL (Sheepridge)
This name is first recorded in the estate survey of 1716, where it referred to a small cottage, so the 'hall' was probably a humorous title. The first tenant we know of was Joshua Stannige and the building he occupied had a ground area of only one rood eleven perches. It was described as lying 'oth west end of Sheepridge'. For over two centuries Bracken Hall remained an isolated

settlement but all that ended just before the Second World War when property in the Northgate area of the town was demolished, and the families moved to a new estate at Bracken Hall.

BRADLEY
This is a common place-name and means simply a broad clearing in woodland. It was formerly an independent estate, certainly at the time of the Domesday survey in 1086, but was later incorporated into Huddersfield where it remained with only 'hamlet' status. That change is likely to have taken place in the 13th century, as a result of depopulation when it was held by Fountains Abbey. In the past it was often called Nether Bradley, in contrast to Over Bradley in Stainland, probably an initiative of the powerful Saviles who had interests in both places. When the Pilkington family acquired the lordship of Bradley, having been abbey tenants in the 1400s, they managed it as a woodland estate, supplying bark, coppice wood and timber locally. They sold the estate in 1829.

BRADLEY GATE
When the Bradley estate was sold in 1829 it was said to lie 'entirely within a Ring Fence' and 19th century OS maps show this fence, the farm called Bradley Gate and extensive woodlands on either side. The farm is almost certainly named after a gate in the fence at that point, described in 1521 as 'a gate called Bradley Yatte'. There may already have been a house there, since 'William Brooke of Bradlaygaite' is mentioned in the will of Thomas Stapleton in 1525.

BRADLEY GRANGE, BRADLEY HALL, BRADLEY HALL FARM
There has been confusion over the exact site of the former grange of Fountains Abbey, partly because of two farm names, that is Grange Farm and Bradley Hall Farm. However, the detailed estate map of 1829 clearly shows Bradley Hall at the end of the ancient highway called Steep Lane. The site now seems remote but when Bradley had independent status this was a prime location, in an open stretch of the Calder Valley, close to important routes and river transport. The 6 inch OS map, surveyed 1848-50, shows the course of the Lancashire and Yorkshire Railway lying directly over the site of Bradley Hall, with just a small outbuilding remaining. Confusingly, on this and subsequent maps up to 1894, a further three separate properties are named Bradley Hall.

BRADLEY LANE (Bradley/Fixby)
An interesting episode in the history of this lane, one section of which was in dispute between Rastrick and Fixby in the 17th century, has already been written about by W.B. Crump. In fact the lane is mentioned frequently in Quarter Sessions records, usually because it was in need of repair, and parts of it were confirmed in 1704-5 to be the responsibility of the Bradley tenants.

BRADLEY LANE (New North Road)
This was the name of a lane that formerly ran from the top end of Huddersfield to Highfields, en route to Marsh and Lindley. In fact, what survives of it is now called Highfields, but once it has passed in front of Highfield Chapel it ends

abruptly at the Halifax Road. Before that road was built the old lane continued along what is now Mountjoy Road and, from an early description of it, we know that it was narrow and tortuous, set among green fields, with a 'high-level paved footway'. It is first mentioned in the court rolls in 1731, and took its name from the Bradley family who were then living near by at New House. The name remained in use into the 1880s at least.

BRADLEY MILLS (Dalton)

These mills mark the site of a fulling-mill with a long history, best known for its association with the Atkinson family from c.1740. The earliest tenant may have been Thomas Kilner who was the constable for Dalton in 1551 and whose family are commemorated in the place-name Kilner Bank. However, Walter Bradley was the miller there in 1688 and Henry Bradley in 1716, a long enough association for the place-name to stabilise. Joseph Atkinson, the first of that family to operate the mill, was said by the diarist John Turner to be of 'Bradley Miln' in 1755.

BRADLEY SPOUT (John William Street)

In Yorkshire the word 'spout' was used regularly for a spring or well, certainly from the 16th century, and it gave rise to a number of minor place-names. It is difficult to say exactly how old a name Bradley Spout is, or what might have been its connection with the Bradley family, but it is recorded on the map of 1818 and was a major source of water in the town over a long period. Later, when people reminisced about the 'old days', they spoke of fetching water from Bradley Spout and it had clearly been a busy locality. In 1826, when a petition was conveyed to Sir John Ramsden asking for 'an abundant and never failing supply of pure water', it stated that during the summer there had been 'on average, at all hours of the day and till late at night, upwards of ten persons' waiting to fill their cans. T.W. Woodhead (1939), revealed that the original spring had been located in a field used for the Railway Station, but the water was conveyed to a new site in the railway wall in John William Street. This was later sealed off but excavations in 1937 uncovered the conduit that had carried the water from its original source.

BRADLEY STREET

This little-known street, between King Street and the Ring Road, actually has a long history, for it is listed in Baines's directory of 1822. It owes its name to 'Bradley, the grocer', who had his premises near by. In 1849, improvements to the area 'surrounded by King Street, Cross Church St. and Kirkgate' were being discussed, and Bradley Street was described as 'the only through communication'. See Broad Tenter.

BRADLEY WOODS

In its Latin form this name is in a Fountains Abbey charter of before 1177 and it is recorded as 'Bradelay Wode' in a Fixby deed of 1290. It was usual then for each township to have an area of settlement and cultivation closely surrounded by trees, much like an island surrounded by water. The encircling wood was given the name of the township but such names fell out of use as the woods were progressively cleared, under pressure from a growing population. Fountains Abbey and their successors the Pilkington family managed 'Bradley Wood' as a springwood estate until 1829 and the name has remained in use, despite more recent changes to the landscape. The *Woodman* near by recalls this association. When the estate was sold in 1829 the woods still covered almost 400 acres, the largest areas being North Wood and Park Wood.

BRAY'S FIELD, BRAY'S WOOD (Deighton)
This was formerly part of Dalton but a change in the line of the river separated it from Kirkheaton parish. However, the boundary stones are clearly marked on some early OS maps, behind the *Peacock Inn* on Leeds Road. The Brays were freeholders, and John Bray was described as a boatman when he died in 1780, so he may have operated a ferry across the river. Certainly his inventory included 'two old boats'. Fields close by had the names 'Broad Ford' and 'Steps Close' so perhaps this was a traditional crossing point, used by Dalton farmers when they needed access to their 'Huddersfield' lands.

BRICK BUILDINGS (New Street)
This was the colloquial name for the row of stuccoed buildings on the west side of New Street, between Lloyds Bank and Cloth Hall Street. They are shown on the town centre map of 1778 and were said by D.F.E. Sykes to have been 'built out of bricks that ... were not needed for the erection of the Cloth Hall'. They were also known as the New Buildings and it is this name that is found on 18th century maps. They were built as superior dwelling houses and shops and, at an initial rent of £20 a year, were thought of as expensive. Brick was an 'alien' material in Huddersfield in the 1700s and the first buildings in which it was used were directly influenced by the Ramsdens. They were familiar with brick in more sophisticated places such as York, Leeds and Wakefield, and no doubt they thought it added a touch of class to Huddersfield.

BRIDGE END
In 1796 there was a 'cottage at Huddersfield Bridge end occupied by John Coats as a public house': this referred to the bridge across the river Colne that we now call Somerset Bridge and the 'cottage' was a public house called the *Star.*

BRIDGE FIELD HOUSE (Leeds Road)
Bridge Field House owed its name to the bridge that carries the Leeds Road over the canal, just beyond where Beaumont Street now is, and it was clearly a villa residence in the mid-1800s. It is difficult now to imagine how pleasant the location was at that time. The house lay next to Oatlands Cottage, on the west bank of the canal, and the two properties had large gardens stretching down to the water. These were decorated with statues and a sun dial, and a boat house had been built on a small inlet. William Learoyd, junior, was living there in 1853, a successful woollen merchant with premises on Leeds Road. He was still there in 1879, with Mr Edwardes of Oatlands Cottage as his neighbour but both

families must have moved soon afterwards. Their properties were acquired by the Corporation, and the new tenant Thomas Dearnley had the backing of some councillors in his attempt to obtain a licence for the house, serving refreshments 'for the convenience of persons resorting to the markets'. The issue was hotly debated in council and finally rejected by those who thought 'it was too paltry a thing for the Corporation to want to earn a few pounds out of a licensed house'.

BRIDGE STREET (Lockwood)

This uncomplicated name draws our attention to the river crossing, less obvious now to motorists than it was formerly to pedestrians and horsemen, and certainly less dangerous. In 1653, for example, the bridge 'was utterly caste downe and overthrowne ... by great inundacion and violence of the water'. Other records tell of the river sweeping people away and gouging out a new course, thereby separating farmers from their fields. In April 1848, the Lockwood diarist Joe William Wilson noted that after a thunderstorm 'the streets were all one mass of water', and people returning from church and chapel had to pay a 'penny a piece to get over in cabs and carts'.

BROAD TENTER (King Street)

The tenter was a wooden frame on which cloth was stretched, after being milled, and it could be dismantled when not in use. By the end of the 18th century, an area of flat land near Seed Hill had been set aside as a tenter ground, to the east of modern Southgate, and it was divided into 'gates' for the convenience of the increasingly important cloth dressers. Earlier it had been customary to erect tenters on uncultivated land, like the one that Richard Williamson had in 1686, standing 'upon the wast in Huddersfeild'. The house called Broad Tenter, first recorded in 1593, stood close to the bottom end of today's King Street, probably near to where there was a tenter frame. It may have been one for broad cloth rather than the more usual narrow cloth. As late as 1838 the Commissioners ordered that a lamp 'at Broad Tenter Field corner' should be removed to Bradley's corner. John Eastwood lived there in the 1590s and Mr William Fillans was the tenant in 1879, just before it was demolished. Subsequently it became the site of a 'yard' that survived until 1959 when it was replaced by the Methodist Mission. It is now part of Kingsgate.

BROCKHOLE BANK (Sheepridge)

This was the original name of the farm called Intake, which is dealt with later. It was located on a hillside close to Woodhouse, at the western end of the Riddings and probably referred initially to a badger set. It can be compared with Brockholes in Honley, although the very first reference to the name, in the difficult court roll of 1532, may say 'Brokhalhbank'. If that reading is accurate it would mean that the element is 'halh' not 'hole' – that is a nook of land.

BROOK STREET

This was 'an intended new street' in 1850, part of the development that followed the demolition of the *George Hotel* and the opening up of John William Street.

When completed it ran from John William Street to Northgate, where it met the older Union Street. In a town where Brook is a very popular surname it ought to be difficult to identify the person after whom it was named. However, the most likely candidate is Joseph Brook of Greenhead Hall who was the chairman of the Commissioners in that year.

BROOKS YARD

There were formerly several yards with this name. One was off Westgate, another off Albion Street and a third off Market Street. Brooks Yard in Westgate was next to the *Cherry Tree Inn*, on the site of the present Estate Buildings, but Brooks Yard off Market Street has survived. In 1879 most of the premises there were occupied by men involved in the woollen industry but there was also a lithographer and a rope and twine maker.

BRUNSWICK STREET

When Brunswick Street was named, in 1859, Sir John William Ramsden particularly asked that 'both sides should be lettered Brunswick Street', perhaps because of the exchanges that had taken place over the name with the Paving Committee. The former New Connexion Chapel was built there but demolished in 1971 when the inner Ring Road was built. What remains of the street is a

Brunswick Street Chapel, built in 1859. After closure it was used by Pickfords, but then demolished to make way for the Ring Road. Courtesy of the *Huddersfield Examiner*

short cul-de-sac, as an extension of New North Parade, and Little Brunswick Street beyond the Ring Road. These names commemorate Brunswick House which was a large villa residence in the mid 1800s, just to the east of Highfield Sunday School. It was the home of a merchant called William Mallinson who had business premises in the town. Brunswick was a popular street name in the 19th century, possibly in imitation of London's Brunswick Square, or more directly as a compliment to the House of Hanover for whom Brunswick was a second capital. Brunswick Place on New North Road was changed to Belmont Street in 1881. See Firth Street.

BULSTRODE
John William Ramsden married Guendolen Seymour, daughter of the 12th Duke of Somerset, in 1865. Her family home was Bulstrode Park near Gerrards Cross in Buckinghamshire. A large house at Bulstrode was built by Judge Jeffreys in 1686. The property passed to Hans William Bentinck in 1706 and he later became the Earl of Portland. One of his descendants, the 3rd Duke of Portland became Prime Minister in 1783 and 1807. The property next passed to the Duke of Somerset in 1810 and the house was rebuilt by 1865. On the death of the 12th Duke in 1885, Bulstrode passed to Lady Guendolen Ramsden, the estate remaining in the Ramsden family until 1932. Bulstrode Buildings, on the corner of Byram Street, thus connects both Ramsden homes. The view from the West Door of the Parish Church, perhaps symbolises the Ramsdens' control of Huddersfield to the departing congregation.

BUXTON ROAD
In 1768 an Act of Parliament led to the building of a turnpike road from Huddersfield to Woodhead, improving trading possibilities for merchants in Derbyshire, Cheshire and the West Riding. The trustees included men from Leeds and Macclesfield as well as several from Huddersfield and the Holme Valley. The route from Huddersfield ran south from the Market Place, on the line of what we now call New Street, and the extension across the Back Green was soon being called the Buxton Road. This name is recorded in trade directories from 1816 and on maps from 1826. The map of 1820 actually says 'To Buxton' on the part of the road beyond High Street. In 1821, a local man recorded the death of 'Old John Attersley, gardiner, (of) Buxton Road, Lockwood'. John Hanson could remember it in those early years when 'all was open fields and hedgerows from the bottom of High Street to the top of Chapel Hill'. When the road was being macadamised, in 1859, the stone for it was 'obtained from the Buxton Lime Company'.

BUXTON ROAD CHAPEL
See Chapel Hill and Underbank.

BYRAM STREET
Byram Hall, near Ferrybridge, became the home of the Ramsden family in the 17th century. It was built for them by the famous Yorkshire architect John Carr and its location and climate made it preferable to Longley Hall. Having moved

The Victoria Temperance Hall in Buxton Road c.1900: the Temperance Society moved to a new hall in Princess Street in 1901. Clifford Stephenson collection

there they were infrequent visitors to Huddersfield, the estate that provided them with a comfortable income. Nevertheless, they were pleased to transfer the name of their new home to several important developments in the town - Byram Street, Byram Arcade and Byram Buildings. Byram Street is on the site of the Swan Yard, demolished in 1879, although the estate correspondence for 1849 reveals that an earlier plan was to give that name to what we now call Lord Street. There was a councillor with the surname Byram, in the later 1800s, but he probably had nothing to do with the street names.

CALTON STREET (Bradford Road)

The housing in this part of the town is of great interest, for it preserves the lay-out of residential streets conceived over 130 years ago. The terraces that run east from Calton Street demonstrate that the term had by then acquired its modern meaning, and their names – Laburnum, Mulberry, Orange, Rose, Hawthorne, Holly and Laurel reflect the uniformity of the building plan. In 1879 many of the residents merited the title of 'Mr', and others had small businesses, so this was evidently not a working class development. Calton Street itself grew out of Calton Terrace, and late additions were Prospect Place (1886) and Daisy Cottage (1887). The present street has several distinctive sections and many of the houses have attractive lintels and decorative work under the eaves. The origin of 'Calton' is not clear, although there is a village of that name near Malham and there were families called Calton in Bradford, Leeds and Wakefield.

CANAL STREET (Leeds Road)

This is a short street that is not shown on the OS map of 1854. It runs between

Calton Street, built in the last quarter of the 19th century for middle class families. The apparent uniformity is misleading, as closer examination reveals.   G Redmonds

Mulberry Terrace, 2007: this is just one of several attractive terraces that lie opposite Calton Street but are set at a different angle.   G Redmonds

Leeds Road and the canal and the name needs no interpretation. It was in 1774 that an Act of Parliament granted the Ramsdens the right to build a canal from Cooper Bridge to King's Mill, and work on the project was completed in 1780. It came to be known as the Ramsden Canal and gave rise indirectly to names such as Dock Street and Quay Street. I have been unable to locate a street of the same name listed in Pigot's early directories, but it seems likely to have been close to the centre of the town. In 1814-15, for example, John Collinson of 'Canal-street' was a boat builder, probably where Dock Street was later.

CANKER LANE (Red Doles)

The word 'canker' is found in several minor place-names and it refers to polluted water. In 1779, for example, a tenant had the rent of his farm lowered because of damage to his land from 'the canker'd or rusted yellow water'. At one time these 'iron streams', as they were also called locally, were commonplace, especially where there had been coal mines, but many have now recovered and people are less familiar with the phenomenon. This Canker Lane is off Red Doles Lane, not too far from Canker Dike which is shown on the 1854 OS map. The stream was reported to the council in 1881 for its 'offensive condition'. As it was referred to in a follow-up report as 'Blackhouse or Cankerwell Lane-dyke', it seems to have been an alternative name for Blackhouse Brook or Dike. See also Ochre Hole.

CARR HOUSE

The earliest record of this name is in 1814, and later maps show that it was formerly on Northgate, close to where the Mechanics' Institution now is. My first thought was that the name might have been coined for a villa residence, but the property seems to have been smaller in 1826 and it lay close to where William Carr had property in 1778, so it was probably named after him or a member of his family. Maps and directories reveal that in the 1850s it was a large house with gardens, the home of Thomas Kilner, gentleman. As building in the 'New Town' gathered pace the land around Carr House was earmarked for development.

CARR PIT ROAD (Dalton)

In the early 1500s there was a farm here tenanted by William Brook. It was an area where floods frequently occurred, sometimes changing the course of the river and leaving deposits of rich alluvial soil. The 'carr' refers to the soft riverside ground, and the 'pits' are likely to have been places where tenants dug out the 'black earth' in order to improve their own land. See Mold Green.

CASTLEGATE

The earliest references to Castlegate are in trade directories, from 1811, although the name appears to have alternated in that early period with Castle Street. The street was in the Low Green part of the town and its name seems to have been chosen quite arbitrarily, inspired perhaps by Castlegate in York, or by Huddersfield's close historic links with Pontefract Castle. We are indebted to John Hanson for a more colourful explanation. It was, he wrote, called after Towser Castle, the name given to the town's lock-up. Towser was the name of the constable's dog. When the Ring Road was built the name was transferred across town to the western section of the new road. We even have a Castlegate Slip that gives access to the town from the New North Road. What remained of Castlegate became today's Watergate.

CAWSEY

This is the original spelling of 'causeway' and it derives from the French word for a paved way, usually where there was a danger of flooding. In this case the name referred to part of the lane that ran from Kirkgate down to the corn mill at Shorefoot. There is a reference in 1620 to a place called the Cawsey but this may be to a house that took its name from the 'causeway'. The meaning is most explicit in entries in the court rolls: in 1642, for example, Huddersfield tenants were condemned for their failure to repair the upper end of the pavement called 'le Cawsey' and in 1690 they were forbidden to fasten their horses between the gates of the vicarage 'and the Causey lane leadeing downe the Streete'.

CEMETERY ROAD

By c.1850 it was estimated that the graveyard adjoining St Peter's Church contained the remains of no fewer than 38,000 former parishioners. Because it could accommodate no more corpses, and complaints were being made about the nauseating stench in the town centre, the Board of Health was asked to investigate. Letters passed between the Huddersfield and London agents of the Ramsden family, and the Improvement Commissioners also became involved. Secret boring tests were carried out on potential locations, and The Huddersfield Burial Ground Act was passed in 1852. Within a year a site at Highfield, near Edgerton, was being 'neatly enclosed and tastefully laid out and planted', and the new cemetery was opened in 1855. Cemetery Road seems to have been in existence by 1876 when it is mentioned in the Improvement Act but it may still have been awaiting development. In 1881, for example, the Burial Ground Committee had its attention drawn to 'the bad state of the road leading from the New North-road to the Cemetery', but no name was given.

The memorial to Joshua Hobson, erected in recognition of the part he had played in securing the new cemetery. G Redmonds

The cemetery at Edgerton, in a picturesque location, with one chapel for Anglicans and a second for Dissenters. G Redmonds

CHADWICK FOLD

Chadwick is a Lancashire surname but it features prominently in Huddersfield records from the early 1600s. The rental for 1797 shows that Thomas Chadwick had property in the town that included Chadwick Croft, along with two cottages and six dwellings inhabited by tenants. They were called Smith, Jepson, Milnes, Bottomley, Kilner and Brooksbank. Chadwick Fold is mentioned from 1811 and is worth noting, as 'fold' was rarely used in this sense in Huddersfield, although common enough in neighbouring places. It was at the bottom end of Kirkgate, on the east side of today's Oldgate. In 1853 there was also a Chadwick's Yard off Northgate.

CHAMBER

We know very little about Huddersfield's town centre in the 16th century, so it seems worth commenting on this rare term. In 1543 the parsonage lands were being sold and among the items listed were one toft, various yards and fields, and several dwellings. Included were two cottages, one called Sykehouse, a tenement in the holding of Thomas Hemingway, and 'the chamber ... adjacent to the cemetery'. This was accommodation for the chaplain of the chantry dedicated to the Blessed Mary - probably a cottage. A family called Sykes was said to be 'of Chamber' in the parish registers as late as 1654 but this was not necessarily the same house. The word 'chamber' normally referred to an upper-storey room but in Yorkshire it could also denote a small dwelling. In 1597, for example, there was 'a dwelling howse called new chamber'.

CHANCERY LANE

John Dobson was a grocer, hop merchant and banker, with premises in the Market Place. When his bank failed, in 1825, the site 'was thrown into Chancery' and this gave rise to the street name. That, at least, is the explanation offered by the local historian, D.F.E. Sykes. I cannot say whether it is true or not, but in any case the name was almost certainly 'borrowed' from London's Chancery Lane, recorded in more or less its modern form as early as 1454. Previously it had been New Street (1185) and 'Chauncelereslane' (1338), having its origins in the office of the Master of the Rolls of Chancery. Originally, Huddersfield's Chancery Lane led out of the west side of the Market Place and then turned abruptly left to join Cloth Hall Street. When it became a through way from Cloth Hall Street into Westgate, in 1862, the new section replaced Lad Lane. In early directories, Chancery Lane had a Market Place address. Chancery Close is the name given now to the old access from New Street. The name had featured earlier on the Huddersfield map of c.1780, along the bottom end of Kirkgate, but this is an isolated example and it may have been put there for its prestige value.

CHAPEL HILL, CHAPEL STREET

These are self-explanatory names but the associations are now less obvious, since the second chapel on the site, the Buxton Road Chapel, was demolished in the 1950s. The original brick building was erected by the Wesleyans on land sold to William Brook in 1775, and it was sometimes referred to as the 'Old

Bank' Chapel. It is shown on the estate map of 1778 as 'Dissenters Chapel' and this description, as well as its position on a freehold site outside the town, reflects the Ramsdens' attitude towards the Methodists.

CHARLES STREET

This street is first referred to in the Paving Committee records in 1848, when a sewer was laid down, and the context makes it clear that it was really an 'intended' street. Levelling followed in 1849 and the earth work was carried out the next year, by William Story. George Brier of Kirkheaton supplied 'setts' to the Commissioners and did the mason work. I thought initially that 'Charles' referred to John Charles Ramsden but a Charles Ramsden worked alongside the Trustees in the 1840s so he is the probable source of the name. See John Street.

CHARTERS GARDENS

These nurseries or market gardens were close to Turnbridge, on land now occupied in part by St Andrew's Road, and they were so called on the OS map of 1854. They take their name from a gardener called William Charters who first appears in the Ramsdens' rent book in 1798. I have been told that he came from Elland, although the surname often has a Scottish origin. His rent was substantially decreased after 1830, and a few years later his tenure was taken up by George Raisbeck. In 1853, the Post Office directory lists this man as a seedsman with a shop in Kirkgate, and the manufacturer Read Holliday was at 'Chartre Gardens'.

CHERRY TREE CORNER

The *Cherry Tree Inn* was a hostelry much used by clothiers in the early 19th century. It stood on the lower corner of Railway Street and Westgate and was demolished in 1868 to make way for the prestigious Estate Buildings. Photographs of it at that time, with groups and individuals carefully posed for the camera, are evidence of its popularity. A new *Cherry Tree* was built but sited diagonally across Westgate, on the top side of the junction with Market Street. This public house was itself pulled down in 1931 when Market Street was widened, although its location is still remembered in the popular name Cherry Tree Corner.

CHINATOWN (Leeds Road)

Peter Walley wrote an interesting article on this name in Vol. 9 of the journal of the local family history society. According to him it was the popular name for the streets called Town Avenue, Town Place, Town Crescent and Town Terrace, although no Chinese people lived there. He wrote:

> *Further down Leeds Road, other estates were known as Tel Aviv and the Holy City, and when we were sat on the wooden railings at Canker Lane, collecting locomotive engine numbers, we could gaze upwards to the hill opposite and see Chatty Nook estate.*

CHURCH STREET

This short street runs between John William Street and Byram Street and it was planned in 1853 as part of the New Town development. Originally it was even shorter and stopped at Back John William Street – now known as Wood Street. In the directory of 1879 the only entry for Church Street was for Mellor & Crowther, plumbers. The street's one claim to our attention is Somerset Buildings, on the north side: this was the town's first library and art gallery, opened by Lady Guendolen Ramsden in April, 1898.

CINDERFIELD DYKE (Deighton)

Few will know this name but it has a long and interesting history. The locality was in Deighton and by the later 1800s the name referred to a place on the Leeds Road, roughly where Couford Grove now is. In 1879 a manufacturing chemist called John Thomas Howarth lived there. The dyke or stream is clearly shown on the canal map of 1773 but 'Cinderfield' is more complex. In 1630, for example, Thomas Hirst of Bradley was ordered at the manor court to repair the way from 'Sinderfore Yate alonge towards Colnebrigge', and Thomas Brooke of Deighton Hall was ordered to clean out the ditches in 'Sinderfore Lane'. In fact there are numerous references to the name over the next two centuries, with spellings such as Cynderfore Dyke (1739), Sinderford Dyke (1763), Cinderfall Beck (1801) and Sinder Ford (1829). We can therefore be reasonably certain that the suffix 'ford' (pronounced 'fore' in the local dialect) gave way to 'field' quite late, which suggests that the lane originally ran from Colne Bridge to Deighton and that there was a ford over the stream near Couford Grove. The cinders would have come from Colne Bridge iron forge and may have been used to surface the lane. Part of the lane near the forge is called Forge Lane on the map of 1773.

CLARA STREET (Bradford Road)

The interest in this residential street, with its neighbours Honoria Street and Eleanor Street, lies in the fact that they were built on land that formed part of the Thornhill estate and were all named after members of the Thornhill family. Thornhill Cottage is close by. William Capel Clarke had changed his surname to Thornhill when he married Clara, the heiress to the Fixby estate, and he had a second wife called Honoria; Eleanor was the name of one of his daughters. The housing development followed on the estate act of 1852 which had enabled the family to grant long-term leases and still exercise some control over the quality of the buildings. In 'The Social Geography of Victorian Huddersfield', Richard Dennis commented on the gardens and forecourts of the houses and their low walls and railings, all indicative of the 'polite respectability of the development'. One section of Clara Street was called Victoria Terrace and the former Cooperative Buildings on Eleanor Street still have their name carved on the entry to the old stable yard.

CLARE HILL

The Collegiate School was founded as a Church of England school in 1838, on a site to the west of St John's Road. It was a rival, therefore, to the College but proved less resilient financially. The first Principal was the Rev. Matthew Wilkinson, a fellow of Clare College, Cambridge, and that connection is likely to lie behind the name Clare Hill. This was first noted in 1844 and it seems then to have inspired the villa names Clare House, Claremont and Clare Bank, which were all listed in that area in the directory of 1879. At that time the Rev. Abraham Smith lived at Collegiate House and Mr C.W. Learoyd at The Headlands. It is tempting to link Clare Hill with Cambridge Road but this name may have originated much later. In the Act of 1876 it is described as 'newly opened out' and its inclusion in the directory of 1879 lists just four names in Claremont Terrace.

CLAY BUTTS (Birkby)

The modern housing estate in the Grimescar valley, close to where the farm Storth was, has taken from us an ancient landscape but it has preserved a link with the past in the street name Clay Butts. We find the word 'clay' in several names in that part of the township, notably in Clay Spring near Cowcliffe where there was a case of sheep stealing in 1697. Further up the valley there is a feeder stream of Grimescar Dike called Tommy Clay Clough, and this seems certain to commemorate a member of the Clay family, who had been resident in Rastrick and Lindley from the 14th century. On the other hand there were formerly two houses in Birkby, one called Clayes (1625) and the other Clay Pits. If the first of these was where John Clay had lived, the second seems more likely to be a place where clay had once been extracted. The latter house is referred to consistently from 1588 when Ellice Roberts was the tenant. In 1604 it was described in a lease as 'all that messuage commonlie called the Claypitt', bought from John Blackburn of Clough House. If Clay Butts was once a field name, before the houses were built, it could be a reference to either the family or the industry.

Thornhill Cottage, Bradford Road: the word 'cottage'
acquired a new meaning as the town's wealthy residents
built themselves expensive houses. G Redmonds

CLOTH HALL STREET

The Cloth Hall was one of Huddersfield's major buildings for over 150 years. The diarist John Turner noted that work started on 20 March, 1765, and the hall was completed the following year. The road that we now call Cloth Hall Street features on the estate map of 1778, although it had no name. There were fields on both sides and just a few buildings at the future junction with New Street – there called South Street. The street names had not yet stabilised, and along today's Market Street were the words Cloth Hall Road. There were alterations to the Hall in 1780, and again in 1848 and 1864, but nothing to fundamentally alter its appearance. Many people were later to find it quite forbidding and it was finally demolished in 1930 to make way for a cinema. The site is now occupied by a supermarket and shops, and all that survives of the Cloth Hall are the street name and a few relics in the grounds of the Tolson Memorial Museum.

CLOUGH HOUSE, CLOUGH HOUSE MILLS (Bay Hall)

Numerous pictures survive of the Clough House, a fine 17th century building that formerly stood on Halifax Old Road, at its junction with Cowcliffe Hill. It is mentioned in several title deeds from the early 1500s, first when Roger Blakeborn was the tenant. Later, the Hirsts and Brooks lived there and Captain Pickering tells us that in 1657 Richard Hirst 'of Cloehouse' was accused of wounding Thomas Brook 'soe that he is in danger of death'. It was the home of Abraham Firth in the middle years of the 18th century. The 'clough' was probably a much older name for the Grimescar valley where 'a corne milne ... in the occupation of Elizabeth Brooke' is mentioned in 1622-24.

COLNE BRIDGE

A charter of 1175-85 refers to Fountains Abbey's bridge over the Colne between Bradley and Heaton, and it was probably on or very near the site of the present 18th century bridge. There was a medieval corn mill close by and then an iron forge and an inn. The settlement grew from the 1500s although many of the houses were on the Kirkheaton side of the river, not in Huddersfield.

COLNE ROAD, COLNE TERRACE

A road of this name was in existence by 1854, close to the bridge in Aspley. It was probably what we now call Colne Street which is mentioned in the Commissioners' minutes in December 1848. The road that we now know as Colne Road was rather later and is first referred to in 1859. In that year it was described as being 'nearly parallel to Firth Street and Aspley Street', and it began 'at the wooden bridge near to King's Mill'. It terminated at Folly Hall, and essential work was still being carried out there in the 1860s. See Queen Street South.

COLNE WATER

From c.1500 this was the usual name locally for the river Colne, and Latin versions of the name take its history back to the 12th century. It was used for the stretch of the river from the Calder to Huddersfield and, formerly, for the feeder stream that we now call the river Holme. For example, a deed of 1236-58, which relates to Armitage Bridge, mentions fields 'juxta aquam que vocatur Kolne' and, as late as 1775, a bridge that linked Cartworth and Upperthong was said at the Quarter Sessions to cross the Colne. It was only after the building of the canal that the river from Marsden to Huddersfield came to be known as the Colne.

COMMERCIAL INN

Anby Beatson who had previously had a grocer's shop in High Street, became the first landlord of the *Commercial* in 1828. He was followed there by John Gill, Ben Ainley and Ben Hutchinson, and the public house they managed fronted onto New Street. A shop occupied the corner site and, by 1874 or just afterwards, this was a tobacconist's run by a man called Wade. It was still a tobacconist's when the *Commercial* took over the premises, and memorabilia in one of the rooms preserve that long-standing link.

COMMERCIAL STREET (The University Campus)

Commercial Street has had a chequered history, and the rather isolated portion of it that survives lies between the University and Firth Street, crossing the canal en route. It must have been exactly the sort of name that the Ramsden family favoured, conveying a clear message of the town's faith in trade. Successive maps and directories emphasise how it developed, starting with Commercial Buildings on the map of 1826. Later there was a Commercial Place, a Commercial Crescent and a Commercial Square, the latter with gardens stretching down to the canal. In the later 1800s it was a residential area with a number of small businesses and lodging houses. In fact the street had formerly been called Clegg Lane, and John Hanson could remember when there were no houses there at all. It must originally have been a footpath, linking the Back Green with Priest Royd, and the earlier name may go back to the 1630s when the Clegg family lived at Broad Tenter.

COOPER BRIDGE

This name has no direct connection with the Cooper family, although they were prominent in Huddersfield for centuries. On the other hand, the fact that their surname was well known locally may have influenced the later spellings. We do not know exactly how ancient the site is. There was a suit in the early 1300s that had to do with a bridge over the Calder somewhere in this area, but no name was given to it. However, Fountains Abbey claimed that they had built the bridge for the convenience of their grange, back in the reign of Henry III, so as to link Bradley and Hartshead. The earliest direct references to the place-name are in 15th century wills when local men left money for its maintenance and repair. In 1426 we find, translating from the Latin, 'the bridge of Cowford' and, in 1442, the vernacular 'Cowforthbrig'. From these it seems reasonable to assume that there was once a ford here for cows and it is worth noting that 'Couford' is mentioned in a Hartshead document of 1336. As late as 1770 a deed refers to land 'near Cooper Bridge within the close called Cowfford' in Hartshead. The change in the place-name is best evidenced in a Quarter Sessions document of 1674 where it appears as 'Cowfurth alias Cowper Bridge'.

COPLEY STONES (Fartown)

There was formerly a house of this name at the foot of Woodhouse Hill. The earliest references are from 1561 when George Horsfall was living there, but the name may be much older, derived from the surname of John de Copelay who was taxed in Huddersfield in 1379. Fields that were part of the property were called 'Copley Mores' (1704) and not too far away were the 'Copply Cliffs' (1716). These all refer to features of the landscape so perhaps the names go back to clearances made by the Copley family. The farm was still there in the 19th century and I am uncertain when it was demolished.

CORN MARKET

This was the name of an area just east of the parish church, reserved for the sale of corn. We cannot be certain how long it had served that purpose but the market is mentioned in the court roll of 1735 when some people were indicted for trading illegally in public houses. In 1778 the market area was a small, irregularly shaped space, open to Kirkgate but with buildings on the other three sides. Two of these were public houses, the *Blackamoor* and the *Masons Arms*, and there were also cloth-dressing shops, a barn, stables, a fold, a pig-sty and a dung yard. The site and the name were lost when Lord Street was built.

COWCLIFFE, COWCLIFFE CROSS

This was originally the name of a hilly and extensive area of common that lay between Huddersfield and Fixby. The element 'cliffe' is a clear reference to the topography but the prefix is not at all straightforward. It certainly has nothing to do with 'cows', for it was spelt 'Cawkecliffe' in c.1570 and 'Caukcliffe' as late as 1769. These seem to point to the Old English word for chalk but that would be inaccurate geologically. Strangely, there is a place-name with almost identical spellings in Newstead in Nottinghamshire, dating from c.1300, and in this case too there is no trace of chalk. It is ironic that the Byron family, who held the manor of Huddersfield until the 1570s, had purchased Newstead Priory in 1540, and that it later became their chief residence. It is just possible, I suppose, that the name was transferred from one place to the other but it seems unlikely, and we are still left with the problem of how it should be interpreted. As Cowcliffe was the site of early quarries, perhaps the whiteness of the sandstone gave rise to the name, either mistakenly or metaphorically. The base of a cross survives on Cowcliffe Hill Road, no doubt marking the ancient cross-roads. It explains the names North and South Cross Roads.

COW MARKET See Beast Market.

CROFT HEAD

This was originally the name of a house and it was located close to the site now occupied by Standard House, a building that separates Half Moon Street and Sergeantson Street. It stood at the head of the Bone Croft and, when Edward Hirste was living there in 1597, appears to have been an isolated dwelling. It is possible that he was the man of that name who had enclosed an acre of land 'from the overgrene', some twenty years earlier. The place-name survived into the 19th century at least. In 1821, for example, 'the road from Market Street to Croft Head, being a public thoroughfare' was provided with gas lamps; in 1850, when it was paved, it was described as a 'yard' near Fox Street. One of Isaac Hordern's photographs shows the house shortly before it was demolished in 1871. The present building was erected on land belonging to a family called Marshall and it was originally known as Marshall's Buildings. Benjamin Marshall moved to the United States in the late 1700s and became a very influential man, operating woollen mills and helping to found a transatlantic shipping line.

CROFT HOUSE LANE (Marsh)
The Croft House estate was put up for sale in 1887 and the catalogue shows that it had been an important villa residence. The accommodation included an entrance hall, a bathroom, six bedrooms and three reception rooms; there was a four-stall stable, a carriage house, a harness room and various ancillary buildings; the grounds covered several acres and were judged to be 'eligible sites for ... villa residences'. The mansion may have been divided between two families in 1853, those of Thomas Brook, a solicitor, and David Sykes, a merchant, later to become a J.P. He was still living there in 1879, by which time there were already numerous houses on Croft House Lane. The tone was set by Isabella Hall who had 'a ladies' school' and Elizabeth Crosland, a vocalist.

CROPPER(S) ROW
Cloth-dressing became well established in the town in the 1700s, and croppers were later at the heart of the local Luddite disputes. They were cloth shearers, the workmen who took the nap off the cloth in the finishing process. This row of houses is mentioned on a gravestone as early as 1807 when it was in a rather isolated position in the New Town. The site has been lost in the development of the roads leading north out of the town, but it was roughly where Oxford Street and Northgate now meet.

CROSS CHURCH STREET
This name is recorded from 1809. The street was part of the expansion that took place to the south of Kirkgate in the opening years of the 19th century, a planned grid pattern that included both King Street and Queen Street. To the south was the Back Green, soon to become Ramsden Street. By 1810, therefore, much of the former Bone Croft had been built on, obliterating the rural scene that old John Hanson remembered in 1878. The *Ramsdens' Arms*, with the Cleggs as landlords, was one of the first buildings on the street and its location was occasionally given as Kirkgate. The street is aptly named, for it 'crosses' from the church to King Street and by a happy coincidence, or thoughtful planning, it also permits the pedestrian a view of St Paul's Church to the south.

CROSS GROVE STREET (Springwood)
When this street was built, possibly around 1840, it 'crossed', or linked, Swallow Street and Grove Street. It is shown on the OS map of 1854, and the minutes for 1851 describe it as being in a neglected state. It lay in the area that is now used as a car park, south of Merton Street, and little beyond the name has

survived. I suppose the enthusiast might discover some of the old 'setts' when the car park surface is disturbed. Grove Street lost its independence in this development, being joined to what remained of Spring Grove Street.

CUCKOLD'S CLOUGH (Fartown)

Cuckold's Clough was once the name of the steeply-cut valley on the east side of Bradford Road, close to the *Ash Brow*. It was known by this name into the late 19th century at least but I have no evidence that it is still in use. In fact, it seems to have given way to Ash Brow in the last 200 years, a transformation that may be hinted at in the Enclosure Award of 1789. Ashbrow Road was described there as a 'public bridle, pack and driving road ... leading across a valley called Cuckolds Clough'. However, when James Mellor was living there, in 1543, the name of his house was 'Cokewald Cloughe', so it probably goes back to the 13th century when Robert Cokewald held property in Huddersfield. This family is referred to frequently in Sandal, from 1297, and their surname seems likely to derive from the word 'cuckold', used derisively of the husband of an unfaithful wife. The spellings are the same as those for the place-name Coxwold, but if that were the source of the surname we should expect forms such as 'de Cokewald', and these do not occur.

DADDLE BECKS (Aspley)

On the map of 1797 this is the name of a field between the canal and the river, immediately north of the tail goit of Shorefoot Mill. Other spellings in that century are Doddle Becks (1765) and Dadle Becks (1716). At first 'beck' seems an appropriate enough element in this location, but it is clear from 16th and 17th century material that it had originally been 'bight', and once referred to a significant bend or bends in the river. The spellings for that period include Doddle Bighte (1631) and Doddilbight (1545). We need also to be careful with the first element which appears to be 'doddle', the local word for a pollard. That would fit the location but a possible alternative is found in the first reference that I have located, in a deed dated 1503. This lists several enclosures in Aspley with 'bight' as the final element, among them 'dubylbyght' or two bends. Although the river in that part of Huddersfield is straight enough now, it had numerous meanders in earlier times, and two of its 'bights' are clearly shown on the map of 1716.

DALE STREET (Chapel Hill)

There is a short entry off the west side of Chapel Hill that bears this name but it is a mere remnant of the original Dale Street. This was a terrace of about a dozen houses, located between Manchester Road and the canal, and I find it referred to in the Commissioners' minutes in 1848. Just south of the entry was the *Grey Horse Inn* and Yard with Miles Dyson as the landlord in 1837. In 1986 the name of the inn was changed to the *Rat and Ratchet*. However, this house may not have been the original inn, for the sign for the Grey Horse Yard survives on an older building just down the hill.

DEAD WATERS (St Thomas's Road)

Dead Waters is a name that has long intrigued Huddersfield people, and the debate about what it might mean returns from time to time to the pages of the local newspaper. We shall not find it on contemporary maps and yet there should be no mystery surrounding its origin since it is well documented and mapped. It referred to an area of low-lying land in the angle where the Colne and Holme rivers come together and where flooding once occurred regularly. It is easy to imagine that Dead Waters described the stretches of water that remained when the floods had subsided. The actual confluence was known as the 'Meeting of Waters' and both these names are on the map of 1768. Under the Improvement Act of 1848 the Commissioners established a depot at Dead Waters where the manure that accumulated in the town's streets was stored and then sold off, presumably to gardeners. The Act of 1876 (Section 22) contains provision for an extension to St Thomas's Road that would start from its junction with Hope Street, pass 'through lands ... called Dead Waters', and terminate at the southern end of Longroyd Bridge.

DEIGHTON

The present spelling of this place-name has stabilised relatively recently, and typical early forms are 'Dictune' and 'Dighton'. The transition is apparent in a land transaction of 1604 where the hamlet is named as 'Deighton alias Dicton'. It is the spelling 'Dictune' that identifies the two elements, the first meaning 'ditch or dike', and the second meaning almost certainly 'farmstead'. The name occurs frequently in documents from the 12th century and is one of the oldest and most thought-provoking place-names in Huddersfield. The first point to consider is that there are at least four other places in Yorkshire with the same name and similar historic spellings. This frequency may even imply that the two elements together had acquired a generic sense – that 'dike-ton' had become a meaningful word in its own right. That may be the case in the place-name Deightonby, a locality near Thurnscoe with the Scandinavian suffix 'by', itself indicative of settlement. Partly because the suffix 'ton' is so often of ancient origin and partly because all the other Deightons are recorded in Domesday Book, it is tempting to think that Deighton in Huddersfield may also have had its origin in the Old English period. If that were the case, it is conceivable that Deighton and Edgerton were originally independent settlements but had become part of Huddersfield before the Norman Conquest. That subordinate status would explain why neither was recorded in 1086.

Most of the evidence suggests that Deighton in Huddersfield was a name with two values. In its primary sense it referred to the settlement which may have been a single dwelling as late as the early 1500s. This was almost certainly where the 'de Dightons' had lived for generations, until they moved into the neighbouring parish of Kirkheaton some time before 1379. They are likely to have been succeeded by a family called Brook who continued as tenants into the late 1600s at least. In its second sense Deighton was a territory which included localities such as Deighton Brook and Deighton Field. We have no map showing exactly what the boundaries may have been, but a land transaction of 1560 provides us with one or two clues. The sale concerned

SUNNY MEAD
DEIGHTON

the manor of 'Dyghton als Dyton, four messuages and three cottages with lands and twelve acres of land covered with water'. It is uncertain where these dwellings were located but families called Hirst, Stead and Jagger were all said to be of Deighton at that time and there may already have been a nucleus of buildings on the original settlement site. The twelve acres under water were almost certainly the 'carrs' or floated meadows alongside the river. By the 1700s Deighton was occasionally being referred to at the Quarter Sessions as the vill or village of Deighton, further evidence perhaps that it was developing into a nucleated community.

DEIGHTON HALL
For much of its history Deighton was territorially independent of Huddersfield, but it only came to be described as a manor at the very beginning of the 16th century. I can find no evidence that the estate ever operated as a manor but the term had certainly come into use by 1504, when the property was owned by the Mirfield family. In 1461, when Oliver Mirfield made his will, he called Deighton a town not a manor, so the change may have been in name only. The Mirfields were important locally, indeed they had been Huddersfield's wealthiest family in 1379, so perhaps they considered 'manor of Deighton' a more prestigious title. It was certainly a description that continued to be used in land transactions of the 16th century, especially when Deighton passed into the possession of the powerful Wentworths. I suspect that 'Deighton Hall' was another name coined to reflect the owners' status. I first find it used in the parish register in 1599 when Edward Brook was said to be 'of Dighton Hall'. In earlier documents he was simply Edward Brook 'of Dighton'. Thereafter the place-name occurs at regular intervals. In 1640, for example, Thomas Brooke 'de Dightonhall' was ordered to 'scower his ditche soe farr as his landes do extend'. In 1798, after further changes in ownership, the land tax for 'Deighton Hall' was paid by Thomas Thornhill of Fixby.

The only evidence we have now for Deighton Hall is the place-name: no trace of a hall or even a substantial old house has survived, although we can see from the OS map of 1854 that the site of the hall was at the top end of what is now called Cherry Nook Road. Whitacre Street and Whitacre Close are near by and they remind us that the Whitacre family of Woodhouse purchased a substantial estate in Deighton in 1786. See Whitacre Lodge.

DEIGHTON LANE
The road through Deighton was formerly part of the ancient highway between Huddersfield and Leeds, but it lost that status in 1765 when the new turnpike road was built through the valley below. One of the earliest references to the old highway is in the Quarter Sessions records of 1641, when the inhabitants of Birstall were held responsible for repairs to their section of 'the king's highway leading between Hothersfeild and Dighton as far as the market town of Leeds'. In the 18th century the stretch through Deighton was called Deighton Lane. In 1712, for example, 'Dighton Lane' was said to be so ruinous that no horses, no carts and not even pedestrians could pass that way.

Sunny Mead, Deighton, in the early 1900s. Rustic names were chosen for the houses that were replacing the green fields. G Redmonds collection

DENTON LANE (Town Bottom)

This lane was at the bottom end of the town, on a site now occupied in part by the Ring Road. It is clearly shown on the map of c.1780, alongside Shutts and Shitten Lane, evidently one of the poorest quarters of Huddersfield. These alleys were probably among those described in White's Directory of 1837 as in need of improvement, 'a reproach on the inhabitants'. Nevertheless, a part of Denton Lane survived and several families were still living there in 1879. These included a chimney sweep, a sack maker, a coal dealer and a pawn broker. The Improvement Act of 1890 gave authority for the lane to be widened but its days were numbered, and no trace of it has survived. It owed its name to the Denton family who lived there in the 1700s. In 1716, Joseph Denton was a blacksmith at the Bottom of Town and James Denton was living there in 1768.

DOCK STREET

Huddersfield's dock yard is shown on the map of 1818, just to the south of Turnbridge. In 1807-8 Timothy Grafton paid the Dry Dock rental, and in 1811-12 John Collinson was the tenant. He was the 'boat builder' listed under Canal Street in 1814-15, so this name may have given way to Dock Street soon afterwards. It is shown on the maps of 1820 and 1826.

DUKE STREET

The name Duke Street is referred to several times from the 1830s but we know little about its early history. It ran between Swallow Street and Grove Street, an area that now forms part of the car park to the west of the Ring Road. Where the name Duke Street occurs in other towns and cities it is usually possible to link it with a particular Duke, but I cannot do that here. Perhaps it was seen as complementing the earlier name Granby Street, or even King Street and Queen Street. In any case it is unlikely to have had any connection with the Duke Street listed in Pigot's Directory in 1816-17, which seems to have been somewhere in the Back Green area. It was possibly a temporary name associated with the *Duke of York* public house.

DUNDAS STREET

This is one of several names that bear testimony to the influence of Lady Isabella Ramsden. She was born in 1790, the daughter of Thomas, the 1st Lord Dundas, and married to John Charles Ramsden in May 1814. She survived his premature death in 1836 and that of his father in 1839. Her son John William was then a minor, and the estate was run by the Trustees until he came of age in 1853. She is credited with having exerted control over the trustees during that period, when Huddersfield was expanding commercially. The street appears as 'Dundas' Street in the Paving Committee minutes of 1849, and the inverted commas may mean that it had still not been officially named. However, it is on the town centre map of 1850. From the start it was largely given over to warehousing for woollen merchants and in 1862 Jonathan Sutcliffe was given permission to erect a temporary shed on the pavement there – specifically to dress the stone needed for the warehouses. No fewer than fourteen manufacturers and merchants had premises on the street in 1879. The Dundas Arms are impaled with those of Ramsden on the Ramsden Estate Building.

DYKE END, DYKE END ROAD (New North Road)

Dyke End was the name of a farm tenanted in 1625 by Edward Symson. It was situated on the south side of New North Road, opposite today's Belgrave Terrace, and is shown on the Jefferys map of 1772. Other tenants of Dyke End in the 1700s were the Brooks and Hirsts. In 1821, when the Commissioners of the Lighting Act were establishing their area of responsibility, that is within a radius of 1200 yards from the cross in the Market Place, the measurements showed that it extended to a point '46 yards this side Mr Midwood's gate stoop, Dikeend Lane'. The lane ran between Trinity Street and New North Road and it was later given the name Portland Street - a change that illustrates the town's transformation in the 1800s.

EAST PARADE (Buxton Road)

East Parade is first mentioned in the minutes in 1847: it extended eastwards to Queen Street South from what was then Buxton Road, more or less opposite the bottom end of South Parade. The residents in 1879 included James Gledhill, a music teacher, and William Gledhill, a pianoforte maker, but there were also machine makers, yarn spinners and woollen waste dealers. The street was demolished when the Ring Road was built.

EDGE HOUSE (Paddock)

This is the name of a substantial villa in Paddock that still overlooks the Colne Valley. In the 19th century it was at the heart of a long and bitter legal case, known as the tenant-right dispute. The house had been built by a man called Joseph Thornton, probably in the period 1835-37 under the tenant-right system, but the landlord, Sir John William Ramsden, gave Thornton notice to quit in November 1861. It is from affidavits that were used as evidence in the ensuing court case that we learn so much about Paddock and Edge House. Joseph Thornton told of how he had selected the site because of the view and the pure air, but it was convenient also, since the mill that he managed was in the valley below. The name he gave it simply described its position – it was he said 'situated at the edge of a very steep declivity' and the gardens were laid out 'on shelvings of broken rock and waste land'. Paddock had been an undeveloped area of stone quarries and common grazing in 1835, 'partly heath or moorland, without soil', but by the time of the trial Edge House had 'assumed a very ornamental character' and Joseph Thornton was not disposed to give it up easily. Its survival is testimony to the case that he and others made in court.

EDGERTON

Edgerton is a settlement site of great antiquity, although until the 19th century it consisted of little more than a single, important dwelling house. It is first referred to in 1311, but it is said to mean the farmstead of 'Ecgheard', a Germanic personal name, and if that is correct then it had a much earlier origin. Indeed, it is possible that the two 'tons', Deighton and Edgerton, were both established long before the Norman Conquest. It was probably where the family called 'de Edgerton' lived in the early 14th century but they were followed by a family called

Cowper or Cooper, possibly from as early as the 1360s. The last member of this family to occupy the house may have been the John Cooper who died in 1677. We have a sequence of family names associated with Edgerton through the 1700s but little other information until the early 1800s when the district started to be thought of as a desirable place to live. It was favoured by the growing middle class, and a street directory of 1837 notes the presence there of 'many handsome villas'. Their names are on the OS map of 1851, and they include Edgerton Cottage, Edgerton Grove and Edgerton Lodge. When the Clarke-Thornhill family sold off land in that area its popularity increased and Edgerton subsequently developed as a fashionable residential area, favoured by successful merchants and manufacturers. The development there linked up with that along the New North Road.

ENGINE, ENGINE BRIDGE (Chapel Hill)

On 19th century maps the bridge over the river at Folly Hall was called Engine Bridge, and a description of it featured in a reminiscence published in the local paper. In 1825-26, according to the writer, it was 'a narrow stone bridge that would only allow of one cart passing over at a time; there were recesses on each side for passengers to shelter in if a cart were passing over'. The origin of the name was explained by T.W. Woodhead when he wrote of the waterworks on the site, built for the town in 1743 by Sir John Ramsden. Water was drawn from the river and passed through wooden pipes to a storage reservoir at the top of the town. The pumping engine that made this possible was powered by a water wheel and it was this engine that gave the district its distinctive name. The 'engine for raising water' was farmed through the late 1700s at a rent of five guineas, and families like the Crowthers and Eastwoods were closely involved. After the turnpike road 'to Buxton' was built, in 1768, Engine Bridge became a focus for further development. It started with a dwelling known as Ingeon House, the home in 1788 of Matthew Lucas and Job Bentley or Bincliffe, but in just over fifty years numerous industrial premises became established there. In 1854 these included a foundry, a dyeworks and several mills, notably Folly Hall Mills. John Hanson's memories of the old engine, as he knew it in the early 1800s, are worth repeating:

> *Whilst a lad, I had to go to Lockwood for milk, as we had no milk hawkers in those days. As I passed Folly Hall I used to be attracted by the screeching and groaning of the old pumping engine. It sounded as if it had not had a drop of oil for twelve months or more. I would peep in through the broken window and watch the crazy thing at work. It would make a desperate effort, stand for a few seconds and then groaningly move off again. Thus painfully and laboriously was the scanty supply of water pumped up from the polluted river.*

FARTOWN

I believe this to be one of the most significant place-names in the parish, even though it is first recorded as late as 1538, in a title deed that refers to 'one dole of meadow in Fyrtowneynge'. Nevertheless, early settlement sites such

The former Fartown Grammar School, located at the north-west corner of Fartown Bar. It was in use as a workshop until demolished in 1971. Clifford Stephenson collection

as Cuckold's Clough and Copley Stones confirm the existence of a 'Fartown' community from at least the 13th century. It is the description of these settlements as the 'far town' that raises important questions.

It is usual to say that the place-name described that part of Huddersfield that lay to the north of the town brook, literally the far part of the town, and that is probably the correct etymology. However, in a number of Yorkshire parishes there are place-names where the element 'town' identifies the site of a pre-Conquest settlement. In Barnsley and Wadsworth, for example, there are places called Old Town and in Methley there is Mickletown.

I believe that Fartown may be such a name, and suggest that it is the site of the first Anglian settlement, the place they named Huddersfield, possibly as early as the 7th century. In that case the nucleus around St Peter's Church would date from just after the Conquest, when Huddersfield became a parish and acquired its first church. If I am right, the decision was made to build it some distance away from 'Fartown', in a new location on the south side of the brook that bisects the township. That would have been more convenient for the parish as a whole, particularly those people living in the Colne Valley.

This 'town brook' is mentioned as early as 1297 and is the key to what might have happened. The rivulet divides Huddersfield from Elland in the upper reaches of the Grimescar valley and then runs directly past the manor house at Bay Hall before disappearing into culverts that channel it into the river Colne. It is insignificant now but it was formerly a major obstacle to both residents and travellers. Place-names such as Hayford and Hebble Bridge remind us of that important crossing.

There are several reasons for supporting the idea of a new site for the church. For example the move would help to explain why Huddersfield is separated from such important structures as the manor house at Bay Hall and the suggested Norman motte near by; it would place the original settlement in a direct line with Deighton, Edgerton and Gledholt, alongside the major highway through the district. That would make sense of settlers moving into the Colne Valley via the crossing of the Calder at Cooper Bridge.

More convincingly it would make sense of the highway system that formerly linked Huddersfield with its neighbours. It has always puzzled me that the old highway from Huddersfield to Halifax ignored possible routes through Lindley or Birchencliffe, both in the same parish, and chose instead to leave the town via the Old Leeds Road and make an abrupt turn westwards to Hill House. If Fartown was the original settlement then Hill House was the logical starting point for the highway to Halifax over Cowcliffe.

'Fartown' would have suffered an immediate loss of status but the area on the north side of the brook continued to be important. The manor house and many of the numerous scattered farms were tenanted by branches of the rapidly expanding Brook family. In 1379 no fewer than four families bore the name Bythebroke and Brooks dominated the total population of Huddersfield in the Tudor period - surely a very significant surname.

FELL GREAVE, FELL GROVE (Bradley)

These two farms were sited close together, just within the ring fence that divided Sheepridge and Bradley. Fell Greave was much the earlier of the two and it has survived, in an oasis of green fields between the two parts of Fell Greave Wood. As 'greave' means wood, or copse, the farm was probably named when the original clearance was made. We cannot be certain when that was but a place called 'Felgreaf' was mentioned in the will of Thomas Brooke of New House in 1554. At that time the farm was probably occupied by one of his kinsmen, for there were Brooks as tenants well into the 17th century. They were tanners, like so many of the family. However, by the early 1800s, the property belonged to a gentleman called Mr Joseph Chadwick, and he was probably responsible for building the neighbouring house named Fell Grove, some time before 1837. The two place-names have almost identical etymologies, since 'grove' too means copse, but Mr Chadwick is likely to have chosen the word because it was becoming a fashionable name for a gentleman's residence. It is the prefix 'fell' that is difficult to explain, although one suggestion by Professor Smith was that it derives from an Old Norse word for a plank or board. He also suggested the surname Fell but I can find no evidence to support that theory. It is worth mentioning though that a place named 'Felege' is referred to in a Fixby deed of 1331. Fixby is a name of Old Norse origin so it may be that Felege was a reference to the steep hill or 'fell' on which the present farm is located.

FENTON SQUARE (Longroyd Bridge)

This is still an attractive group of houses and it would have been even more pleasant before the locality was so heavily industrialised. The Fenton family, who lived near by at Spring Grove, owned several small but important freeholds in the area and these houses had been put up quite early in the 1800s. Their decision to call the development a 'square' may mean that the houses were originally intended to attract reasonably well-off tenants, and there are notes in a surviving account book that throw some light on that. We learn, for example, that some of the properties had been fitted with out-kitchens and that it was intended to rail off the yard and plant evergreens. There was mention also of the problems these plants might have because of the smoke in the neighbourhood.

It was not the only group of buildings to be given the family's name. There were two terraces called Fenton Row, for example, one just a few hundred yards from Fenton Square, nearer to Outcote Bank, and the other in Shelf, between Halifax and Bradford. There was also Fenton's Court in Halifax, and Fenton Road in Lockwood.

FIELD GATE (Sports Centre)

Early spellings of this name, such as 'Fieldyate' (1639) and 'Feildyate' (1655), make it clear that it referred originally to a gate rather than a highway. The place-name survived into the mid-1800s and is shown on maps, identifying a mill and a dwelling house towards the bottom of Lowerhead Row - behind the present Sports Centre. It is probable that it marked the point where the town formerly gave way to cultivated fields – the beginning of the town lane. The gate may have been there from a much earlier date, but is first recorded in the 1600s as the name of a house on the site. The first families known to have lived there were the Websters and the Hirsts, the latter sometimes described alternatively as 'of Lane'. As the town expanded in the 1800s there was more building in that area and the name fell out of use.

FIELD HOUSE, FIELDHOUSE LANE (Leeds Road)

The farm called Field House owed its name to the open fields that formerly occupied the valley floor below Sheepridge - alongside the present Leeds Road. The settlement is known to have been there in the 1480s, and a surviving title deed records a lease of the property to John and William Brooke in 1493. Other families who lived there were the Horsfalls, Mitchells and Shaws. In 1676, for example, Robert Mitchell of 'Feildhouse' had a yew stolen during the night and Nicholas Singleton was accused of having taken it into the town to be butchered. It remained a rural location into the 19th century when coal, clay and ganister were mined there. The OS map of 1854 shows a colliery, coke ovens and a brick field close to the canal. Old Fieldhouse Lane and Fieldhouse Bridge are names that remind us of the very early settlement.

FIRTH STREET

The history of Firth Street goes back to 1854 when it was proposed to have an outlet sewer 'below the canal' through land owned by Mr Thomas Firth of Toothill in Rastrick. From May 1856, work was carried out on two new streets in that area and the intention was to call them Aspley Street and Firth Street. A minute in the records of the Paving Committee throws light on the influence that the Ramsden family continued to have on such matters:

> In consequence of Sir John William Ramsden exercising the right of naming New Streets within the Limits of the Improvement Area, this Committee respectfully call the attention of (his) agents to the desirability of not having more than one street of the same name.

Apparently Firth Street was already the name of a street near the Infirmary, so Sir John William had this called Little Brunswick Street instead. During the 1860s Firth Street became a focal point for local entrepreneurs, and within a decade their premises lined the new road from Priest Royd Mills to Upper Aspley. There were whitesmiths, machine makers, millwrights, brass founders, cotton spinners, yarn spinners, cloth finishers, woollen manufacturers and fent dealers. Among the new place-names were Melbourne Works, Commercial Mills, Larchfield Mills and Gladstone Mills. Messrs J. Firth and Son had Grove Mills, and they built additional warehouses and offices in the late 1880s. By then the street had assumed a character that is still discernible in the buildings that have survived.

FITZWILLIAM STREET, LOWER FITZWILLIAM STREET

Earl Fitzwilliam was the brother-in-law of Lady Isabella Ramsden, and an influential trustee of the family's estate in the period 1839-53. The street named after him was initially a relatively short one. In 1848, for example, it ran only from Trinity Street to Dyke End Lane (later re-named Portland Street). Not long afterwards it was continued in a direct line right through the northern part of the town, ending at Leeds Road. The central section disappeared in the Ring Road developments but the bottom end survives as Lower Fitzwilliam Street. It is perhaps worth noting that Lady Isabella's mother was Lady Charlotte Fitzwilliam.

FLASH HOUSE (Fartown)

The farm at Flash House was demolished some years ago and the place-name has not survived. The buildings were sited close to Fartown Bar, on rising land to the right hand side of Fartown Green Road, and they are shown on early maps. I first find the farm mentioned in a lease of 1542 where it is described as 'a mease called Floshe House' in the tenure of John Broke. A later John Brook 'of Flash House' was a subscriber to Highfield Chapel in 1772. The meaning is not in doubt, for the specific element refers to a pool or watery place. In 1623, for example, a farmer at Hunsworth, near Birstall, was ordered to release water from 'the floshe' on his land. In the OED there are actually separate entries for the words 'flash', 'flosh' and 'flush', all with much the same meaning

Kershaw House, Fitzwilliam Street: a typically elegant town house, with good ironwork and an attractive stone façade.   G Redmonds

and onomatopoeic in origin, like the word 'splash'. The spellings for this name and those for Flush House in Austonley are very similar and suggest that our ancestors considered the words to be synonyms.

FOLD
A fold was initially a small enclosed piece of land, usually adjoining a dwelling. However, the word was also used locally of a cluster of houses built in such a fold, as early as the 1600s. In this case, though, the name may have referred to a single house, and it occurs several times in the period 1581 to 1664, mostly in the parish registers. Thomas Shaw is the first resident of whom we have a record and he was followed by families called Blackburn, Sykes and Brook. We can only speculate about where Fold was and why it disappeared but the dates suggest that it may have been in the area where the Market Place was created, soon after 1671.

FOLLY HALL (Lockwood), FOLLY ROAD (Cowcliffe)

Several place-names in the Huddersfield area have 'folly' as a first element and there have been at least two places in the township called Folly Hall. In Cowcliffe, for example, there was a Folly Hall as early as 1716. It is not well documented but features on some maps and is commemorated in the name Folly Road. The second Folly Hall is the district around the bottom of Chapel Hill and it is much better known. The traditional explanation, based on comments by John Hanson in 1878, is that the name was given to a house built on that site by Marmaduke Hebden. This was in the late 1700s, in an undeveloped part of the town well away from the centre. It had poor communications and seemed altogether a foolish choice. However, Marmaduke Hebden was a major sponsor of the new road to Woodhead and Buxton (1768), and he owned land in Lockwood at the point where the road crossed the river. His fine new residence, built in the vernacular style, is shown on the plan of 1768.

FOUNTAIN STREET

This street lay between Brook Street and Fitzwilliam Street, in the area now occupied by the Tesco supermarket. It stretched from Northgate to the Bay Hall footpath, before John William Street was built, and the first reference I have to it is in 1834 when it was provided with lamps. The line of the street is actually shown on the map of 1826, with 'Wells' at the Northgate end and 'Spouts' to the east, and these almost certainly explain the name. On the map of 1851 the 'Spout' had become a 'Fountain' and Wells Mill stood next to 'Brooks Wells'. At the eastern end was the Brick Factory, 'owned by Messrs Roberts Bros'.

FOX STREET

George Fox is perhaps the least well-known of the Trustees but his name is remembered, alongside those of Dundas and Sergeantson, in the development that took place around the Cloth Hall c.1848. The site was drained and paved in that year and the map of 1850 shows the street directly to the south of the Cloth Hall, parallel with Dundas Street. There were 'Gardens' here in 1826. The Vagrant Office was then directly behind the Cloth Hall, towards the top end of Fox Street which continued to its junction with Upperhead Row. The name survives but the location has changed: Fox Street now is the short street to the north of where the Cloth Hall stood.

FRIENDLY STREET

This is a little-known street that runs north from Northumberland Street, along the west side of the old Mechanics' Institution. This building was opened in 1859 and has recently been restored. The street name commemorates its later use by the Friendly and Trades Club.

FRIZING MILL (Aspley)

'Frizing' was a finishing process in the manufacture of cloth which involved drawing the woven 'piece' across a rough surface, so as to raise the nap. In west Yorkshire generally this process had been carried out in water-powered mills from the 1720s, and possibly in horse-powered mills from even earlier. In Huddersfield, the Atkinsons of Bradley Mills were called 'frisers' in 1758 and, just ten years later, Mr John Atkinson, senior, was the tenant of the Frizing Mill at Aspley, paying £2 rent annually. His mill was located on an extension of the tail goit of Shorefoot Mill, downstream from Somerset Bridge and between St Andrew's Road and the river. It was probably rather older, for it was said in the rental to have been formerly in the tenure of a man called Marriot.

GEORGE HOTEL

The present *George Hotel* dates only from 1850 when it was built close to the recently completed railway station, the second of two iconic buildings in what was being described as the 'New Town of Huddersfield'. It was the third inn in the town to bear the name, and it replaced an earlier *George* that had stood on the northern side of the Market Place, where John William Street now begins. The town was reluctant to lose this fine Georgian building, judged to have been built in 1787, and it was therefore re-erected at the junction of Byram Street and St Peter's Street. It still occupies that site and there are two date stones high up on its blank, south-facing wall, one for 1787 and the other for 1687, exactly one hundred years earlier. The latter is thought to have come from the first *George,* an inn built to enhance and take advantage of the town's new Market Place. Unfortunately there is no reference to its name at that time but the inspiration was certainly St George and not the first Hanoverian George who came to the throne in 1714. In 1713, for example, William Radcliffe carried out some of his duties as a Justice of Peace in 'a chamber at the sign of the *George* in Huddersfield'.

GEORGE STREET

The expansion of the town to the west of Upperhead Row started with the Waterworks Building (1828) and Spring Street (1829), and the fine town houses that remain show that they were intended for Huddersfield's growing middle class. Work appears to have continued with South Street, mentioned in 1842, and then with a series of new streets that included those between Upperhead Row and the Cloth Hall. The development is best understood if we look at the results fifty or so years later, when a grid plan of wide and spacious streets occupied an area that stretched as far west as today's Park Avenue, and southwards from Trinity Street to Prospect Street. George Street was being levelled and paved in 1849-50, and it ran initially between Upperhead Row and South Street. Later it would be extended in both directions and by 1907 was a continuous line from the Cloth Hall to St Peter's vicarage. Now, a short section survives between Sergeantson Street and Upperhead Row, and a longer section westwards from South Street. This is called Upper George Street although its earlier name had been Springfield Terrace. Ironically, the original George Street disappeared when the Ring Road was built. I do not know which 'George' it commemorates although the date might point to the trustee George Fox.

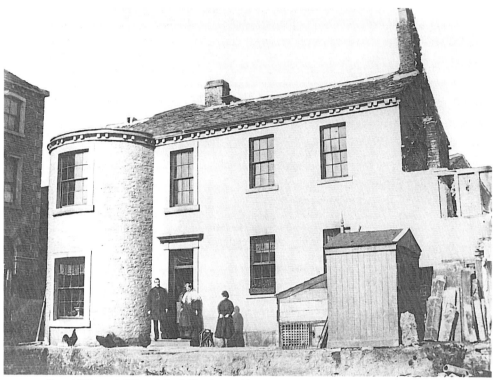

The old Vagrant Office at Croft Head, roughly where Lower George Street now is. It was pulled down in 1869.   Clifford Stephenson collection

GLEDHOLT

Gledholt derives from two words meaning 'kite wood', but it was already a settlement in the 13th century, when a family of that name was living in Huddersfield. One or two of these very early references are not dated, but Sybil de Gledeholt is mentioned in 1274, and John de Gledeholte in 1289. The surname appears to have died out about one hundred years later but we cannot be certain, since it is written 'Gledhill' in some sources. This popular Yorkshire surname is connected with Barkisland and Stainland but it has no obvious place-name source so the apparent confusion with Gledholt raises questions about its origin. In the early 1400s the Nettletons and then the Shaws lived at Gledholt but the rent fell into arrears in 1453 and the estate reverted to its owners, the Mirfields. They leased the property to a Quarmby draper called William Hirst, and members of this family occupied the property for the next 250 years. Alice Hirst of Gledholt sent a petition to the magistrates in 1692 declaring that her rates ought not to be assessed under Huddersfield as the house was in the hamlet of Marsh. More recent tenants have called the elegant mansion there Gledholt Hall but there is no historic justification for that.

GOLDTHORPE'S YARD (King Street)

The entrance to this long-established yard is off King Street and it was much altered in the recent Kingsgate shopping development. Before the renovation it had been in a very poor state of repair, having lost its final residents in the 1980s. Now it serves as a seating area for Yates's wine bar. In 1879 there were four houses or dwellings in the yard, two occupied by Mrs Carver and Mrs Atkinson and a third by Henry and Edward Wear who were plumbers. The fourth was used as a centre by the Huddersfield Naturalists Society, founded in 1848. The buildings seem likely to date from the early 1800s so the yard is probably the one named as Laycocks Yard on the map of 1851. Henry Wear was a plumber there in 1853 and 1864.

GRAFTON PLACE (Commercial Street)

Grafton Place was in Back Commercial Street, just to the east of St Paul's Gardens, where there is now access to the University from Queensgate. It was a short residential street, not shown on the OS map of 1851 but included in the directory of 1853. There was a Grafton Street in London, named after the Duke of Grafton, and that may have been the inspiration for the name.

GRANBY STREET (Upperhead Row)

This was formerly the name of a narrow and mainly residential street off Upperhead Row, which is first shown on the map of 1820. It almost certainly took its name from a public house 'at the Top of Huddersfield ... known by the name of *Granby's Head*' (1771). This had previously been part of a property called the Bowling Green, and it was tenanted by John Hirst whose ancestors had held land there in 1716. The public house commemorated the Marquis of Granby (1721-1770), the Scarborough man who was Commander-in–Chief of the British Army in 1766. Land Tax records for 1799-1805 have entries for John Hirst, '*Marquis*'.

GREENHEAD, GREENHEAD PARK, GREENHEAD ROAD

Greenhead Park preserves the name of an estate that had its origins almost 500 years ago, in a house located at the 'head of the green'. The house stood on its own and was tenanted over the years by the Hirsts, Wilkinsons and Fentons. It is shown on the estate map of John Lister Kaye in 1791, with its expansive gardens, a pond and the dog kennel yard. Greenhead Lane, now Greenhead Road, ran from the house to Gledholt with fields on both sides but it was then a private way between the two properties. The Waterworks Act of 1876 contained provisions for widening and improving the bottom end of the road, near 'the Vicarage House'. The Park was officially opened to the public in 1884 but only after the defeat of a plan to set the land aside for villas. That debate took place in 1869, and it was the dedication and financial support of Alderman Denham that saved Greenhead for the public. Despite considerable opposition he took out a lease on the land and the 'park' was unofficially opened in 1870. The site of the mansion was not included within the grounds of the Park and it was finally demolished in 1909, to make way for Greenhead High School.

GREENHOUSE, GREENHOUSE ROAD (Fartown)

The reference in this case is to Fartown Green, although the name is actually recorded before Fartown itself. The present 'house on the green' was identified by Edward Law as one of the few domestic buildings in Huddersfield designed by Edward Blore, the eminent London architect. In fact the settlement has a much longer history and records show that the site has been in continuous occupation for over 500 years. In 1504, Richard Beaumont of Whitley Hall leased 'Grenehowse' to James Blagburne and it was subsequently tenanted by generations of Brooks. However, when Edmund Horsfall was in occupation, in 1689, the house was registered as a meeting place for Quakers, a move that suggests it may have been used illegally for that purpose during the years of persecution. The licence was renewed by John Horsfall in 1712.

GREENSIDE

This name referred to the 'green' at the top end of the town and it may have been in popular use for longer than the records suggest. In fact, it is first shown on the map of 1820, as an extension of Westgate, but this was soon afterwards given the name West Parade, and Greenside dropped out of use. However, that did not happen immediately, for the names were listed together in the Post Office directory of 1853.

GREEN STREET

Green Street is now somewhat isolated. It was part of the New Town development and in 1852 was one of three new streets 'at Newtown' where the levels were being fixed. The others were Oxford Street and Newtown Row. 'Green' may refer in this case to the surname but that is pure speculation.

GRIMESCAR

This is actually a Fixby place-name, but I include it here because it is popularly used to describe the valley that divides the ancient parishes of Huddersfield and Halifax. Both parts of the name are likely to be Old Norse in origin, with 'scar' referring to the steep slope on the Fixby side of the stream. The first element may have been the personal name that is found in compounds such as Grimthorpe and Grim's Dike, said to derive from a word meaning 'a masked person' or 'one who conceals his identity'. Examples are relatively late. In 1576 a man called Ratcliffe was the tenant 'of Grymescar' and in 1590 'certain Colyers working in Grymskar in Fekisbye' discovered the remains of an ancient kiln, the source of the Roman tiles and pottery found at Slack. A Lindley map of 1609 shows the farm Grimescar Foot, close to the stream. This is of course, the 'town brook' of Huddersfield in its lower reaches.

GROVE STREET

The villa built at Spring Grove by William Fenton is probably responsible for this name. It is not on the earliest town maps but appears among the listed streets from 1837. It gave its name in turn to the *Grove* public house.

HALF MOON STREET

The first reference that I have found to Half Moon Street is on the map of 1825 but it features much later in trade directories. Perhaps the name was added to the map when Thomas Dinsley was dealing with the Lighting and Watching Commissioners in the 1830s. In any case London's Half Moon Street is likely to have been the inspiration, and the Ramsdens would certainly have approved of the name, if indeed they were not responsible for its choice. This London street was apparently the location of an inn called the *Half Moon*, said to be a heraldic alternative to the crescent moon.

HALL, HUDDERSFIELD HALL

This place-name and its location are both poorly documented. Between 1587 and 1632, several men called Armitage were said to be 'of Hall', or 'of the hall in Huddersfield', notably Samuel Armitage who became a wealthy merchant in London. After the Armitages came the Horsfalls and the Jepsons and then, finally, Henry Collingwood 'de Huddersfeild Hall' in 1687. None of these references tell us exactly where the hall was, but in the court roll, just three years later, Henry Collingwood was ordered not to 'stopp the ancient way through his croft called Tinkeler' and this identifies a possible location. In fact, on the map of 1716, Tinker Croft lies opposite Hall Croft, close to where the Field Gate was. Tinker and Tinkler have the same meaning so the hall may have been the detached house at the western end of Tinker Croft. In any case it seems likely that it was in that part of the town.

HAMMONDS YARD (King Street)

This is possibly the most attractive of the surviving yards, and it takes its name from the Hammond family who still had the property in the 1980s. The last

Hammond's Yard in 1976. The back of Queen Street Chapel and St Paul's Church can be seen to the right. Clifford Stephenson collection

owner was Miss Margaret Dixon who kindly showed me the earliest title deed when I visited her in 1982. It was a lease, dated 1807, of land 'lately parcel of the Bone Croft ... staked and marked out to be made use of for erecting a Dwelling house and other buildings'. Provision was made for drains and cellars, and King Street was to 'be enjoyed as a footway'. The lessee on that occasion was Miss Dixon's ancestor William Wilks, the surgeon who played an important part in the building of the town's infirmary. He lived in the yard and there is a reference in the minutes of the Lighting Committee to 'Wilks Passage' in 1821; the Post Office directory of 1853 lists it as Wilks' Yard. The change in the name came after 1840 when William Wilks died; his daughter Susannah had previously married a Kirkgate tea dealer and grocer called Joseph Hammond and they moved into the yard.

HARDY BECK, HARDY BRIDGE (Fartown)
The first record of these two names is in 1778, in an indictment of a highway at the Quarter Sessions. The road was said to run through Fartown 'from the north end of a certain bridge called Hardy Bridge, situate over ... a certain rivulet called Hardy Beck, to the Turnpike Road'. These were alternative names for the town brook and Hebble Bridge and they reflect the local importance of the Hardy family. They lived at Lane in the 18th century and their surname has a long history in and around Huddersfield. However, the first man said to be 'of Lane' was Thomas Hardy in 1714. It is interesting to note the use of 'beck' instead of 'brook', for this element was of Scandinavian origin and more typical in areas further to the north. However, the same change was taking place in other parts of Huddersfield, and 'beck' was later to become the usual word for a stream.

HAT, HATTERS, JOLLY HATTERS (Westgate)
The inn appears to date back to 1763 at least, and the landlord in 1778 was James Armitage. It stood at the top of town, in the network of alleys which bore the names Midwood Row, Marshall Row and the Maze in c.1780. In the 1700s the tradesmen described as 'hatters' were prominent in the commercial life of the town, with three listed in the directory of 1754-58. Since it was not unusual at that time for a beer seller to have his own trade, it is possible that a hatter's establishment gave rise to the name of the inn. The *Olde Hatte* in Trinity Street survives to remind us of the earlier inn.

HAWKE STREET (Union Street)
This is another 'lost' name. It referred to a short street that ran between the eastern ends of Northumberland Street and Union Street. The name occurs in the minutes from 1853 but the evidence suggests that work continued on its development over the next decade. This street and its neighbours were demolished when the alignment of Leeds Road was changed. Old Leeds Road marks the former line. The origin may have something to do with Lord Hawke who had strong local connections.

HAWKSBY'S COURT (New Street)

This is one of the few yards to survive, although it is in a rather sorry state at present. And yet it was once a desirable town centre location: in 1879, for example, it was 'home' to a solicitor, an accountant and the librarian to the Huddersfield Church Institute and School Association. A woollen manufacturer also had premises there and this may be a link with its earlier history. We cannot be certain just when the yard was built, but it was well established by the 1840s. It is shown in its present location on the town centre map of 1851, and the directory of 1845 lists Henry Hawxby, wool stapler, of Hawxby Court. However, the Hawksbys were already wool staplers in the town in 1784 and their links with New Street certainly go back to the rental of 1798. The wool staplers Hawxby and Sutcliffe had a New Street address in 1822, as had John Hawxby in 1837. He was an accountant and share broker. This unusual surname derives from the parish of Haxby, near York, but the family was well established in Leeds from the 17th century and this may be where the Huddersfield Hawksbys came from.

HAYFORD (Fartown)

This was the name of a field immediately to the south of the town brook and it is recorded as 'Hayforth' in 1250. In 1689, a manorial by-law ordered the tenants not to 'make any footeway over two closes of land called Towne Brooke and Heyford' and thereafter the name features on several maps. It presumably marks the location of a ford that was there even before the 'hebble bridge', and it would have been used by the tenants as they brought in their hay from the Town Ings.

HEATON FOLD, HEATON ROAD (Marsh)

Little is known about this location, although Thomas Priestley was living in 'Heaton Fold' in 1744 and the Dysons were tenants there of the Ramsden family for several generations. The place-name survived into the 20th century, for it is shown on OS maps, roughly where Dingle Road now is. The old cottages there probably mark the site and Heaton Road preserves part of the place-name. It is certainly a very old settlement and it was probably the original home of a family called Heaton who lived in Marsh from at least the 1590s. Roger Heaton of Marsh is the first of the name. See Thorpe Hall.

Hebble Street. The former town brook passes almost directly underneath these houses on its way to the river Colne. G Redmonds

HEBBLE BECK, HEBBLE STREET, HEBBLE TERRACE
(Bradford Road and Fartown)

These are fascinating names that somehow evoke the small, medieval township. When Huddersfield was no more than a village it was separated from the 'far town' by the stream that comes down the Grimescar valley and enters the Colne near the Football Stadium. It was then known as the 'town brook' and we can still catch glimpses of it at Bay Hall and on Willow Lane, although much of it is now out of sight, passing under the modern roads and streets through culverts. On the OS map of 1851 it is called Hebble Beck and the houses on the west side of Bradford Road are Hebble Terrace. Hebble Street survives to remind us of those names. The significance of the 'hebble' is that in this part of Yorkshire it is the dialect speakers' term for a narrow bridge. The name Hebble Bridge which is on the same map has a long history.

Hebble House, Bradford Road: it is on the line of the town brook and close to Beck Road, a name derived from the more recent Hardy Beck. G Redmonds

In 1650, for example, the inhabitants of Huddersfield were accused at the manor court of neglecting to repair the highway that ran between the corn mill at Shorefoot and 'le Townebrookehebble'. See also Hardy Beck.

The quality of the stone carving and metal-work on Hebble House is remarkable and demands our attention. It seems not to be well known. G Redmonds

HIGHFIELD(S)

The source of this name was a large enclosure on high ground to the west of the town, recorded as 'Hie Fielde' in 1581. The inference is that it was formerly part of the township's communal arable land, but the evidence is late. However, an 18th century field book lists some of the joint tenants, including Nicholas Bramhald and John Hodgson. We come across one of these surnames in 1748, when it was ordered at the manor court that John Bramhall's neighbours should not 'make any footway over a close called the High Field', possibly through what is now the cemetery. The area began to figure more prominently in Huddersfield's history when the Independent Chapel was built there in 1772. The lane from the chapel into the town at that time was described as a 'deep narrow cart road, accompanied by its high–level paved footway'. All around were green fields. It was later decided to have the municipal cemetery in that part of the town.

HIGH PARK (Bradley)

This was the name of a hamlet on the north side of Bradley Road, just east of the lane leading to Lamb Cote. When the estate was sold, in 1829, families called Ibbotson and Hinchcliffe lived there but the land was being farmed by Thomas Hardy. Bradley Park is referred to in deeds from 1549, and Park Wood lay just to the west of Bradley Hall, a substantial part of a characteristic egg-shaped enclosure that may define the original park. However, High Park is not in that part of Bradley and it appears to be a relatively late name so there may be no direction connection with the park.

HIGH STREET

This is a classic name nationally, used in many towns and cities for the street which is at the very centre of commercial activity – often the principal street. In such cases the word 'high' points to importance and not height, and can therefore be compared with the first element in 'highway'. Huddersfield's High Street has no such history, for the name was chosen arbitrarily, almost certainly because it sounded important. In fact, it was formerly the western section of the lane across the Back Green and it is sketched out on the maps of 1716 and 1778, with fields to the south and just a few buildings close by. In 1878 John Hanson recalled that as a boy he had known two one-decker cottages at the lower end of the street, in an area that 'comprised the whole of Buxton Road and the greater part of High Street'. There was also a footpath that 'ran across the fields to Outcote Bank'. The first record I have of the name is in the directory of 1809-11.

HILL HOUSE, HILLHOUSE LANE (Fartown)

The hill that gave Hill House its name is not particularly high, but it lies on the former highway from Huddersfield to Halifax, close to King Cliff, and on its summit is the mound that is thought to have been a Norman motte. The site is clearly marked as 'Mount' on the OS map of 1854. The hamlet was another of the numerous Fartown settlements with 'house' as the second element and, like most of the others, it was occupied for a long period by a branch of the Brook family. Roger Brooke 'of hilhouse' was the tenant in 1532, and John Brooke was involved in the tithe dispute with the vicar in 1696. According to Philip Ahier the Elizabethan house was demolished in 1871, although pictures of it survive. See Beacon Street.

HONORIA PLACE, HONORIA STREET (Bradford Road)
See Clara Street.

HOPE ON ANCHOR (Shore Head)
Edward Law has traced the history of this public house from 1799, when Edward Haigh was the landlord. He gives the name as *Hope and Anchor*, and traditional explanations of this make the point that the anchor was one of the most ancient symbols in Christianity. However, in Huddersfield the name frequently appeared as *Hope on Anchor* and a possible interpretation of this was offered by Barrie Cox in English Inn and Tavern Names (1994). In the 14th and 15th centuries, for example, it was not unusual to have inn names which employed the phrase 'on the Hoop', such as *George on the Hoop* and *Hind on the Hope*. The preposition 'on' probably indicates that the sign was placed in the hoop of a barrel or on the flat circular end of a suspended barrel. When the allusion was lost the elements in the name could easily have been transposed, helped by the fact that 'a hope anchor' was the name given to a spare anchor.

HORSE AND JOCKEY (Market Place)
The map of 1778 has this inn on the south side of the Market Place, with a brewhouse and stables. Robert Perkin or Parkin was the landlord followed by Thomas Parkin, last recorded in 1816. John Hanson remembered it as 'an old public house called the *Doll Hole*' which might be interpreted pejoratively. 'Doll', a pet form of Dorothy had several colloquial meanings – among them that of a prostitute. The name may be one of the oldest in the town, possibly in existence by the 1740s, and this influences the meaning of 'jockey'. In the 1700s the word was commonly used in the north for a horse dealer, especially one who used sharp practice. The essayist Macauley had this in mind when he referred to 'a Yorkshire jockey'.

HORSE SHOE (Kirkgate)
This was the name of a Kirkgate inn that was finally demolished in 1883, probably when Seth Stringer was the landlord. In his annual review of Huddersfield in 1880, the mayor spoke of 'the old and objectionable buildings in Lower Kirkgate' that had already been cleared away. These were near the *Horse Shoe Inn* and their demolition was the first step towards improvements authorised by the Act of 1876. The earliest reference to the *Hors shoo*s is in the Brewster Sessions of 1777, and the inn is well documented over the next 100 years. Backed by its two extensive stable yards it stood between Rosemary Lane and Denton Lane, a block of buildings that reached almost to Castlegate.

The Horse Shoe Inn, *Kirkgate: the sign bears the name of Seth Stringer who was the landlord when it was demolished in 1881.* Clifford Stephenson collection

HUDDERSFIELD

We do not know exactly when Huddersfield acquired its name but it is of Anglian origin and these settlers are thought to have arrived in the district no earlier than the 7th century. If that is so it is worth noting that something like 400 years were to pass before the name was recorded in the Domesday Book. There it is spelt 'Oderesfelt' and 'Odresfeld', a word that is clearly made up of two elements. The suffix or generic 'feld' can be recognised as the word that gave us our modern 'field' but we should not use the modern word to explain the place-name, for 'feld' has changed its meaning subtly over the centuries. Originally it referred to a tract of open country, that is one without trees, but later it meant arable land, and Margaret Gelling and Ann Cole recently suggested that in 'Huddersfield' it may have the latter meaning. They looked more closely at the distribution of the element and concluded that it seems to have been used at the junction of low and high ground. This interpretation may fit Hill House better than it fits the site of the modern town. Locally the name can be compared with Mirfield and Wakefield, both relatively important communities, so it may be that the suffix is also an indication of early local status. The real difficulty is to offer a convincing explanation of the prefix, and here the other early spellings have to be considered, since 'H' was often the initial letter, e.g. Huderesfeld, Huderisfeld and Hoderesfeld. Professor Smith was of the opinion that these derived from an Anglian personal name but he based that theory on the single example 'Huddredisfeld' recorded in 1241, and the suggestion seems less attractive when we learn that no other example of that name has been found. The Swedish scholar Eilert Ekwall compared the place-name with Hothersall in Lancashire but, like Smith, he was relying on a poorly documented personal name. Gelling and Cole have suggested an Old English word 'huder' meaning a 'shelter'. Finally, we should note that typical spellings from the 1500s are Hothersfeild and Huthersfeild, and these represent the dialect pronunciation. See Fartown.

HUDDERSFIELD BRIDGE

This was an earlier name for the bridge that crosses the river Colne to the south of the town, now called Somerset Bridge. The earliest reference that I have located is in the will of William Broke in 1537. He left '6s 8d for the mending of Huddersfield bridge' and seems likely to be the William Brook of Carr Pit in Dalton who is mentioned in the court roll of that period. On Senior's map of Almondbury, dated 1634, the name is written 'Hothersfield Bridg' and there are similar examples in the 17th century court rolls. See also Long Bridge and Somerset Bridge.

HUDDERSFIELD GREEN

Much of what is now the built-up central part of Huddersfield was still uncultivated land in the 16th century, referred to simply as the Green or Huddersfield Green. That must have been the case through the Middle Ages and yet one of the earliest references is in a deed of 1520 which mentions a newly constructed house on the 'greyn'. It was a large area and different sections were later defined more specifically, notably as the Back Green, the

Nether Green and Over Green. Encroachment on these areas of common was for both agricultural purposes and for building and eventually the emphasis was on the latter. Typical of the 16th century encroachments was William Horsfall's modest clearance in c.1570, which consisted of five roods 'inclosed from Huddersfield grene a little from the shorehead'. In 1704 Susan Bedford sent a petition to the magistrates for help, saying that 'being destitute of an Habitacon' she had 'gott leave from Sir William Ramsden ... to build a cottage upon Huthersfield Green, being the wast ground in the said Mannor'. She claimed that she was being prevented from doing so by a freeholder called Abraham Firth. There can have been little unexploited common land left by 1800 but the name Back Green was still in use as late as 1830. In that year the Lighting Commissioners had to order brick-makers who were working on the Back Green to refrain from burning the bricks on site, because of the nuisance this caused to the townspeople. The common way across the Back Green later became Ramsden Street.

IMPERIAL ARCADE (New Street)
The term 'arcade' was borrowed from the Continent where it originally described a vaulted walk way, one with a succession of arches. It became a fashionable word in England in the late 1700s for a shopping centre, and Burlington Arcade in London is possibly the best known of these. Lion Arcade may be the earliest example in Huddersfield, completed in 1853, and it was followed by Imperial Arcade and Byram Arcade. Imperial Arcade takes its name from the *Imperial Hotel* which stood on the east side of New Street in the mid-19th century, directly opposite Hanson's Yard. It is said in <u>The Old Yards of Huddersfield</u> (1986) that the arcade was erected by a Mr J.R. Hopkinson between 1873 and 1875, linking the space in Hanson's Yard with the yard of the *Queen's Hotel* in Market Street.

IMPERIAL HOTEL (New Street)
According to D.F.E. Sykes the *Imperial Hotel* was originally a private dwelling house but it became a hotel in the 1840s and is on the OS map of 1851. An early photograph shows it as *Vickers' Imperial Hotel,* a three-storey building with a dozen columns on the ground floor and an imposing entrance. The two upper storeys are in the vernacular style and they survive unchanged to the north of Woolworth's. There was access from the hotel, via Imperial Yard, to Victoria Lane, and from there, until quite recently, the name could still be read on the back of a tall building – now demolished. Samuel Bradley was the landlord in 1853, when an omnibus used to 'meet the trains'.

IMPERIAL ROAD (Edgerton)
This is the third local name to recall the days of Empire, and it is later than those in the town centre. In a survey of Edgerton, carried out in 1977, the writers noted the variety of housing in Imperial Road, the differing sizes of the gardens, and the range of architectural styles. They commented also on the way the status of the houses seemed to change according to the distance from Halifax Road - from villas and 'middle class' to 'better quality working class'.

These were built over a period of time, from the 1850s into the early part of the 20th century, but the earliest ones were villas. According to Jane Springett, the agents who acted for the Clarke-Thornhill estate exercised control on the location and architectural quality of the villas erected in this part of the town, and they in turn influenced the street names. Among the names of neighbouring villas were Luther House, Hungerford House and Cleveland House, and these all gave rise to street names. Along with Thornhill Road, and several streets in the Fartown area, they remind us of the influence of the Thornhill family.

INTAKE, INTAKE BANK (Sheepridge)

This locality is not shown on modern maps but it can be clearly identified on those of the 19th century. The 'bank' was the steep slope between Sheepridge and Fieldhouse Bridge, and the 'intake' referred to some earlier clearance there. 'Intake' was land that had been 'taken in' from the waste or commons, as opportunity offered. We learn from the Houghtons' Family Memorials (1846) that they had a farmhouse 'at Intake' for generations but that by 1846 it had already been demolished and the ground 'blended with the Woodhouse estate'. The place-name's earlier history is complicated by the fact that Intake had an alias: from 1532 at least it was an alternative name for the Sheepridge farm called Brockhole, held by members of the Brook family. Fortunately this alias is occasionally explicit but its evolution can be traced in a variety of sources. In the early 1600s, for example, Thomas Brook was said to be of Intack or of Brockholebank, depending on the source used. In 1631 the Ramsdens' account books record payment from Thomas Brook 'of Intack for his lands at Brockholebancke' and in 1817 a title deed mentions 'the messuage commonly called Brook Hole Bank otherwise Intack'. As Brockhole Bank was recorded long before Intake the inference is that the name changed after an important clearance, possibly c.1600. See Brockhole Bank.

JACK BRIDGE, JACK HILL (Birkby)

The stream that flows through the Grimescar valley is mentioned frequently in this book, partly because it was an important boundary and partly because it had different names at different times. On the OS map of 1854, the stretch that passed the bottom end of King Cliff Road had the name Hardy Beck, and it was crossed by Jack Bridge, much used by travellers from Fartown on their way to Birkby or up Blacker Lane. The earliest references that I have located are in the accounts of Fartown's surveyors of the highways. In 1747, for example, Luke Clay was paid for some minor work on 'Jack Brigg'. Now, the stream passes under the road through a culvert, on its way to Bay Hall, and the bridge is forgotten. However, the footpath that once ran up the hill to Halifax Old Road has been surfaced and is known as Jack Hill. In field names 'jack' often refers to 'unused land' but here it seems likely to commemorate some long-forgotten resident of Fartown.

JOHNNY MOOR HILL (Paddock)

The association of the Moore family with this part of Paddock goes back to the late 18th century at least. In 1797, for example, John Moore paid 4s rent on a 'newly built' cottage. The family expanded in that area over the next forty years and the census returns of 1841 list no fewer than eighteen people called Moore whose address was given as 'John Moore Hill'. The oldest was Martha, aged 65, a lady of independent means and almost certainly the widow of the John Moore mentioned above. The importance that the family attached to the name John is clear from the fact that three of the males in 1841 were so called, one of them a clothier, the other two cloth dressers. Clearly it was not a poor family and that point is worth making in view of the locality's later reputation. As late as the 1970s some Paddock families had a very poor opinion of Johnny Moore Hill, and that resentment seems to have been linked to a bequest made in the will of William Cliffe in 1850. This made provision for sums of money to be paid to poor widows aged forty and upwards, except, that is, for widows from Johnny Moore Hill who were excluded. Those who administered the charity in the 1970s told how the young 'Johnny Moore Hillers' at Paddock School had been required to put their caps in a separate box, instead of hanging them in the cloakroom - because they would be 'full of nits'! It was David Clarkson who traced the source of that poor reputation, when he wrote about an outbreak of cholera in Johnny Moore Hill in 1849: the disease was not understood at that time but it was linked in the newspapers with people 'of irregular habits, those living in low and filthy localities', so by association Johnny Moore Hill came to be thought of as such a place.

JOHN STREET (Buxton Road)

The Civic Centre and a variety of other public buildings and amenities now occupy the area between High Street and the Ring Road, and little survives of the busy community that lived there from the late 1820s. The street that would be named John Street is on the map of 1826 but unnamed and undeveloped; it cut through the centre of the area from east to west, roughly parallel to High Street. The land to the south was simply a large field, and the open space between Buxton Road and Albion Street was marked as 'building ground'. Work must have started almost immediately, for the Lighting Committee required lamps to be placed in John Street in 1829. By 1850 a network of streets had been created, with shops, offices and small manufacturing enterprises: there were also several yards and courts, including Roebuck's Yard, Jowitt's Square and Fanny Kaye's Yard. All that building had taken place in a very short space of time. John Street was almost certainly given its name in honour of Sir John Ramsden who died in 1839. Subsequent changes in that part of Huddersfield have been just as dramatic, for all that survives now of John Street is Buxton Way.

JOHN WILLIAM STREET

This name commemorates Sir John William Ramsden, MP who inherited the family estate as a minor in 1839 when his grandfather died. However, until 1853, when he came of age, the estate was administered by a group of trustees and it was their agent, Alexander Hathorne, who suggested this street name. The names and titles of the trustees also survive as street names, i.e. Sergeantson, Dundas, Fox, Fitzwilliam and Zetland. Sir John William was to be an active landlord, with his own plans for the 'New Town' of Huddersfield, and his influence would eventually be considerable. However, the major decisions about the next phase of development had already been made and work was certainly being carried out on John William Street in 1850. A letter from Alexander Hathorne, in the Paving Committees accounts for August 1851, captures the moment when the 'gateway' to the New Town was opened. He asked the Commissioners to proceed with the work that would take John William Street 'through the site of the present *George Hotel* into the Market Place'. In many ways the demolition of the old *George Hotel* and the creation of John William Street symbolised the hopes that both the trustees and the Commissioners had for the opening up of Huddersfield. The building of the station had made expansion to the north of Kirkgate and Westgate desirable, and the first new building to be completed was the present *George Hotel,* in 1850. The streets of the New Town were then laid out on a grid pattern and the process of sewering, levelling, draining and paving began.

KING CLIFF ROAD, KING CLIFF FLATS
This a misleading name. The suffix is accurate, for the land drops sharply down to the town brook and Jack Bridge, but 'King' is a relatively late development. In the Fartown surveyors' accounts, for example, Joseph Greenwood was paid in 1767 'for levelling Kiln Cliffe', and in the Enclosure Award of 1789 one of the new roads was said to run 'from Hill House … over Kiln Cliffe Common'. It is uncertain just what sort of kiln was once there.

KING STREET
On the town map of 1826, King Street is a broad and very straight thoroughfare, running from New Street down to Broad Tenter and Shore Head. It was of no great age, although it ran along the line of an earlier footpath, and John Hanson could remember the early 1800s when there were only hedgerows and fields in that part of the town. The earliest references that I have found to the name are both for 1807, the first in an estate rental and the second in the title deeds for Hammond's Yard. The street was clearly in the first stage of its development at that time but progress was rapid and by 1826 there were buildings on both sides of the street, many of which survive. The name should be seen as linked with that of its neighbour, Queen Street, and it was almost certainly a tribute to George III.

KINGSGATE SHOPPING CENTRE
Those responsible for planning the shopping centre doubtless felt that they had given it a historic name, a complementary name to Queensgate and a neat link between King Street and the old 'gates' of the town centre. They either did not know that 'gate' was a regional word for a street, or they chose to ignore that in their desire for something traditional.

KING'S HEAD BUILDINGS (Cloth Hall Street)
The *King's Head* was a prominent public house on the south side of Cloth Hall Street, one that was formerly much used by clothiers attending Huddersfield

Kingsgate, 2007. The name links elements of King St and Old Gate and is a happy combination, although it ignores the local meaning of 'gate'. G Redmonds

Market. The name is recorded from 1803 when John Senior was the landlord, and coaches operated from there to Halifax and Doncaster soon afterwards. The yard associated with the inn stretched as far back as Imperial Arcade and there is an evocative water colour of it by H. Bishop, painted in 1923, just before the *King's Head* was demolished. With its steps and balconies and small individual rooms the picture offers us a real insight into what the yard was once like. Now only the name survives, high up on the buildings that replaced the old inn.

KINGS ARMS
There was a public house with this name on the east side of the Market Place, and Edward Law thought that it might date back to the developments there in 1711. However, the name was first recorded in 1778-79 when Martin Taylor

The mills at Kingsmill Lane: the buildings were damaged by a serious fire in 1967 and demolished in the early 1980s. Clifford Stephenson collection

Blackburn

was the landlord; he was still in business there in 1803. Its history after that is less clear, since there was another *Kings Arms* in Lowerhead Row. Directories for the period 1818-1853 place this inn near Field Gate, with James Bottomley as the landlord, so the probability is that this was the *Kings Head Arms* shown there on the map of 1851.

KING'S MILL, KINGSMILL LANE (Almondbury)

There was a corn mill on the Colne from the Middle Ages but I have found no reference to it as the King's Mill until 1522. It served Almondbury but was also confusingly called Huddersfield Mill on occasion and Queen's Mill when Queen Elizabeth was on the throne. The 'royal' prefixes recognise the Crown's possession of the Honour of Pontefract to which Almondbury belonged. The last corn to be ground at the mill was c.1915, after which textile machinery was installed. However, both the major buildings were subsequently damaged by fire and demolished. The name King's Mill Lane previously referred to a lane that linked the mill with Newsome.

KIRKGATE

For many centuries this was Huddersfield's main street, possibly its only street, and in that early period it was known first as the town gate and then the town street. The fact that Kirkgate derives from two Scandinavian words meaning 'church street' immediately gives it an aura of antiquity but the truth is that in this case it dates back only to c.1797. The estate maps and documents suggest

The Star Inn, *Kingsmill Lane, rebuilt in 1900, and pictured here in 1906. The carter is transporting a warp beam, to or from a woollen mill.*   G Redmonds collection

that Sir John Ramsden even considered calling it Church Street, but there were streets called Kirkgate in most of Huddersfield's more important neighbours and that no doubt influenced his choice. Other traditional names were chosen about the same time so that Huddersfield, at a stroke, was put on a par with Leeds, Wakefield and Bradford. From the Beast Market the street bends to the south on its way towards Shore Head, and this section was sometimes called Old Kirkgate. In the minutes of the Paving Committee, in 1853, they commented on 'the small amount of traffic passing down Kirkgate', an indirect result of the construction of so many other streets. It eventually lost its function as a major route through the town and the lower end became Oldgate.

KIRKMOOR

This is probably one of the oldest of Huddersfield's place-names and it is referred to in a variety of documents from the mid 1500s. At that time it was an enclosure of arable land, probably part of the former town fields: a title deed of 1545 records the conveyance of 'two leie (fallow) landes in the nerre Kirkemore'. As late as 1716 the 'selions' or strips in Kirkmoor were shared out among a number of men, although the original enclosure had by then been sub-divided. A reference in the manor court of 1724 makes that clear, for tenants were ordered not to make any way over or through the five closes ... called Kirkmoor'. These fields lay to the east of Norbar, later renamed Northgate, on either side of today's Lower Fitzwilliam Street. The suffix 'moor' suggests that the land may once have been common grazing and that it had some direct connection with the church, but we have no evidence to throw light on that relationship. We may, though, be able to take Kirkmoor's history back even further, for a charter of 1297 mentions a clearance called 'Kirkthauet', a name that may employ the Scandinavian word 'thwaite'. The use of 'kirk' and the location suggest a connection with Kirkmoor.

KIRK MOOR PLACE, KIRKMOOR STREET

Neither of these street names had a long history. Kirkmoor Street is shown as an 'intended' street on the map of 1850, roughly where Byram Street now is, but it is not there on the map of 1851. It is distinct from Kirkmoor Place which was then a group of houses on the east side of Northgate, close to the bottom end of Fountain Street. In the census returns of 1851 something like thirty families were resident there, many of them from Ireland. Richard Dennis commented on the insanitary conditions of the houses there, noted for their 'foul and offensive drains'. Neither of the names survived very long and the housing finally disappeared when the Ring Road was built. It is almost impossible to visualise it now as you drive under the sign 'Unna Way'.

LAD GREAVE WOOD (Bradley)

This wood lay on the right hand side of Leeds Road, close to Colne Bridge, and it covered the area between the present *Woodman Inn* and the river. It is clearly marked on the sale plan of the Bradley estate in 1829, but not on the OS map of 1854. Although it had survived the building of the canal, it lay right in the path of the new railway line and was probably felled as that was being developed. It may be that 'lad' can be taken at face value, referring to a youth or serving lad, but the word was also used locally for a standing stone and that may be the meaning in this case. A document of 1474 refers to 'Ladgreve' as the property of Fountains Abbey.

LAD LANE (Westgate)
I can find out very little about this name but it applied to a narrow alley that linked Chancery Lane and Westgate, the two joining at 'Mr Haley's warehouse': to the west of the entry into Lad Lane was the inn called the *Swan with Two Necks*. In 1862 plans were approved for a new bank in Westgate for the Halifax and Huddersfield Banking Company, and this involved the demolition of the public house and the Stamp Office. As a result Chancery Lane was extended and straightened and Lad Lane ceased to exist. The old warehouse that survives on the lane may be the one that had belonged to Mr Haley, especially as the owners in 1862 refused to co-operate with the developer. Although I have found the name Lad Lane only from 1837, on maps and in trade directories, the use of 'lane' as the generic may point to its survival from a much earlier date, possibly as an unofficial name. See Chancery Lane.

LAITHE (New House)
This is the dialect word for a barn, but Edward Law has shown that there was a farm of this name, in the late 1600s, on land belonging to the Bradleys of New House.

LAMB COTE (Bradley)
There was a small settlement here, from the 16th century at least, and it was home for generations to the Gibson family. Also 'of Lamb Cote' were the Kayes who had an alehouse there in the 1620s. Two of the earliest references I have noted are 'Lamcoat' (1565) and 'Lambecoyte' (1568), the second of which represents the dialect pronunciation. A 'cote' was an animal shelter or a small dwelling and the first of these meanings seems preferable here. Perhaps it was linked originally with the neighbouring farm of Shepherd Thorne, before it became a farm itself. In any case the two places remind us that Fountains Abbey had large numbers of sheep in Bradley from the 13th century.

LANCASTER'S YARD (Cloth Hall Street)
This yard is on the north side of Cloth Hall Street, opposite the King's Head Buildings, and until recently it was being used by a picture-framing company. At the time of writing the yard is no longer accessible to the public and its fate seems uncertain. Its attraction lies in the fact that it has preserved the features we associate with the yards used by clothiers, similar to those shown in the painting of the King's Head Yard. It is now generally referred to as Cloth Hall Chambers, but was earlier known as Laycock's Yard, after the plumber and electrician who ran his business there. It came to be known as Lancaster's Yard in the 1850s but no connection has been established with the prominent Lancaster family who lived in Queen Street.

LANE, HUDDERSFIELD LANE, LANE HEAD, TOWN LANE (Leeds Road)

At the point where the town gate of Huddersfield reached the open green at the town bottom it forked, with one branch heading towards the bridge over the Colne and towns to the south, and the other branch running down the valley to the north east. This latter highway was formerly the departure point for travellers to Leeds, Bradford and Halifax and it also gave the townspeople access to the town ings. As soon as it had passed the Field Gate it was variously described as the town lane, Huddersfield lane or simply 'the lane'. After several hundred yards, and once the town brook had been crossed, there was an important junction: straight ahead lay the town ings, whereas an abrupt turn to the west took travellers towards Hill House. Even beyond that point the highway may still have been known as the town lane. For example, a Quarter Sessions document of 1654 refers to the 'way to Halifax, in a certayne place called Cawcliffe, in the Towne Laine'. It is certainly an ancient term, with references to 'Huddersfeld layn' in a deed of 1520 and 'the Towne Lone' in a will of 1537. 'Lone' was the regional word for 'lane' and we find this spelling as early as 1379 when Richard Bithelone was a Huddersfield resident. This man's name may be the earliest evidence of a settlement on the line of the lane, possibly at the junction mentioned above. There is a hamlet called Lane there on the map of 1716 and there are numerous references to it from the 16th century. Armitage, Blackburn and Brook are the names of families living at Lane in Elizabeth's reign. The Armitages were said to 'of Lane Head' in the 1600s, so that may have been an alternative name for the junction.

LEAROYD BRIDGE, LEAROYD STREET (Leeds Road)

The Learoyds were prominent woollen manufacturers, and D.F.E. Sykes said of them that they were responsible for 'the town spreading out its arms with fresh vigour to Bradley'. Their Trafalgar Mills certainly made the surname familiar to all those who passed along Leeds Road. However, these mills were built only in 1895-96, and the Learoyds had earlier been established at Lane, where Hillhouse Lane branches off the Leeds Road. They were listed as manufacturers in directories from 1814-15 and figure repeatedly in the Commissioners' minutes. In 1850, for example, Mr Edward Learoyd sought permission 'to lay pipes across the Leeds Road for water from the Canal for his new mill' and just two years later it was resolved to put down a kerb 'from the Canal Bridge at the Lane to the top of the road leading to Mr James Learoyd's mill'. Their name is remembered in Learoyd Street, just off the Leeds Road, but it is not generally known that the canal bridge was formerly known as Learoyd Bridge. In the directory of 1853 Edward Learoyd's home address was actually given as 'Canal Bridge'.

LEEDS ROAD

We associate this name with the direct route from Huddersfield to Cooper Bridge, roughly on the line of the river and the canal, but it formerly passed through Fartown Green, Woodhouse, Sheepridge and Deighton. In 1740, for example, the inhabitants of Fartown were ordered to repair 1000 yards of the highway in the Hill House area. The first road along the valley was authorised by the Turnpike Act of 1765 and it cut a new line through the valuable ings or meadows in the valley bottom. Such Acts did much to popularise the word 'road' and it gradually replaced 'highway' in most records.

LEE HEAD (Birkby)

This was probably the 'head' or upper end of the common called Stony Lee and a family called Brook was living there in the 1560s: Edward Brooke was one of several tenants with that surname who enclosed land from the common in c.1570. The house called Lee Head is not referred to until 1589 but the tenant was John Brook and the evidence suggests that both place-names, that is Stony Lee and Lee Head, refer to the same location. The Brook family remained at Lee Head for generations but other surnames linked with the hamlet are Armitage (1620) and Kitson (1669).

LIME KILN LANE or ROAD (Aspley)

Edward Law traced the history of the Aspley lime kilns back to 1808 when they were operated by Richard Clay. They were located on the south side of the canal, just to the west of the corn mill goit, and there was an access road along the canal side, running in both directions. The line of that road is shown on the map of 1825 although at that date it was not named. In 1837 Joseph Kaye was operating the kilns: he was Huddersfield's most famous builder and no doubt used the lime in his mortar. His interest in lime burning was said by Edward Law to go back to 1821. In 1854 Kaye requested that Lime Kiln Lane should be mended at the town's expense but the Commissioners 'could not entertain the application', so it was clearly not considered to be a public right of way. It is shown in Aspley on the OS map of 1851 as Lime Kiln Road, in that location, with three clusters of kilns and two lime pits. The canal towpath that passes through the University grounds is roughly on the line of the road.

LION CHAMBERS, LION ARCADE

Lion Chambers is one of the first Huddersfield buildings to catch the eye of visitors who arrive by train. On the skyline is the huge lion which gives the building its name although the present statue is only some thirty years old. The former stone lion was removed in January 1977 and sent to Newcastle where it was remodelled in glass fibre. In White's Directory of 1853 the Lion Arcade comprised 'externally, long ranges of elegant shops and warehouses in the Italian style; and internally an extensive arcade or covered market, fitted up with warehouses and stalls for the sale of cloth'.

LITTLE BERMONDSEY (Lost)

Bermondsey is a district south of the river in London, to the east of Tower Bridge, but it is likely to have been the inspiration behind Huddersfield's Little Bermondsey which was part of Temple Street. In 1837 the wool staplers John and Matthew Hirst had their premises there.

LOCKWOOD'S YARD (New Street)

This can be a very confusing place-name for family and local historians. For example, the yard which now runs between New Street and Victoria Lane, down the side of Marks and Spencers, was formerly known as Lockwood's Yard although the name has not been in use for some time. It is linked with the parallel yard that we now call Market Avenue – formerly Greenwood's Yard – and both are clearly shown on the OS map of 1851. Two other yards with the same name were discussed in <u>The Old Yards of Huddersfield</u> (1986), and both may have been associated with Lockwood's Mill on Upperhead Row. In 1849 the first of these was directly opposite the top of Granby Street, next to the mill. These extensive mills were devastated by fire in 1914. There is actually an attractive drawing of the third yard by Noel Spencer which places it 'off Duke Street – between Grove Street and Swallow Street' – close to the rebuilt Lockwood's Mill.

LODGE (Bradley)
This was the name of a farm in Bradley, tenanted for generations by a branch of the Hirst family. James Hurst 'de Lodge' was the tenant in the 1590s and Francis Hirst was in occupation in 1829 when the Pilkingtons sold their Bradley estate. The farm was in the former Bradley Park and doubtless owed its name to the hunting lodge there. It is worth noting that Heaton Lodge, which lies just across the river in Kirkheaton, was named in the 1790s when 'lodge' was becoming a popular name for a gentleman's villa.

LONGROYD BRIDGE, LONGROYD LANE
The 'long royd' that gave rise to this name was an assart or clearance, possibly going back to the 12th century, but it is not recorded independently. The first reference to the bridge is in a title deed of 1501 that conveyed rights in the course of 'Colnewater' to Richard Beaumont and Nicholas Byron. They were joint lords of the manor of Huddersfield and they owned a fulling mill near by at Paddock Foot: those rights extended from 'Drycloughende' to the 'Longroidebrigg'. The family most closely associated with the running of the mill was called Thornton, and William Thornton was said to be 'of Longroid Lane' in 1632. A petition to the magistrates in 1721 claimed that Longroyd Bridge needed to be rebuilt. It was, the petitioners said, 'an antient wood bridge' that linked the market town of Huddersfield with its Yorkshire trading neighbours and also with numerous places in Lancashire, Cheshire, Staffordshire and Wales. Reference was made to the 'frequent passage of packhorses and other carriages'. However, long before Huddersfield became a market town the bridge linked the Colne Valley with the routes to Bradford and Leeds via Gledholt Bank.

LONGWOOD HOUSE (Netheroyd Hill)
This hamlet was just on the Huddersfield side of the boundary with Fixby and its history seems certain to go back to the 15th century. Some writers have mistakenly taken the name to be 'Long Woodhouse', influenced by the two neighbouring hamlets of Woodhouse, one in Rastrick and one in Huddersfield. However, it owes its name to the Longwood family who in turn derived their name from the place called Longwood, some six or seven miles to the south west. The earliest reference to their surname is in 1485, when John Longwodd

was fined at the manor court in Rastrick, but the name of the house is not actually recorded until 1608 when 'the messuage called Longwoodhouse' was granted to Nicholas Thornhill and Edward Hanson. The Longwoods had left the area some time before that, possibly about 1570, but the house was undoubtedly named after them.

LORD STREET

Plans for a street to the east of the churchyard were first discussed in 1849, as part of the New Town development, and it was realised immediately that many well-known landmarks would have to be demolished. The buildings that overlooked the Corn Market were inevitable victims since they occupied the intended junction with Kirkgate. Indeed, the *White Horse*, one of Huddersfield's oldest hostelries, stood on the site where the new street would join Kirkgate. The original intention was to call this new street Byram Street but there was considerable delay before work started and the idea was eventually shelved. As a result Byram Street became the name of the parallel street on the top side of the churchyard and the demolition of the *White Horse* had to wait until 1857. Work on Lord Street started soon afterwards but it was still being described as an 'intended street' in 1865. It may have been named after a family called Lord who had premises in St Peter's Street or, alternatively, Sir John William Ramsden may have chosen it in honour of his father in law, the Duke of Somerset.

LOWERHEAD ROW (Beast Market)

Leeds has an ancient street called the Headrow, first recorded in 1523. I imagine that it initially referred to a row of houses at the head of the town but by the 1600s it was extensive enough to have its western end called the Upper Headrow and its eastern end the Lower Headrow. These names were doubtless approved of by the Ramsdens who transferred them to Huddersfield early in the 1800s, possibly at the same time as they decided on Kirkgate and other historic names and for the same reason. The fact that 'Raw' or 'Row' was already the name of houses in the town may have influenced their choice. However, in both Leeds and Huddersfield the relationship between the different parts of the names was not clearly understood and they were often mistakenly written as Upperhead Row and Lowerhead Row. More influenced by 'upper' and 'lower' than by 'head' the Ramsdens gave the name Lowerhead Row to the street that led out of the lower part of the town, a location which completely separated it from Upperhead Row and paid no respect to its etymology.

LOW GREEN, LOWER GREEN

The name Lower Green had replaced Nether Green by the end of the 17th century, dispensing with what had become a vernacular term. The reference in the Enclosure Award of 1789 to a carriage road 'over the Low Green ... called Low Green Road' illustrates the final stages of its development and, when that road became Castlegate, the name finally disappeared completely.

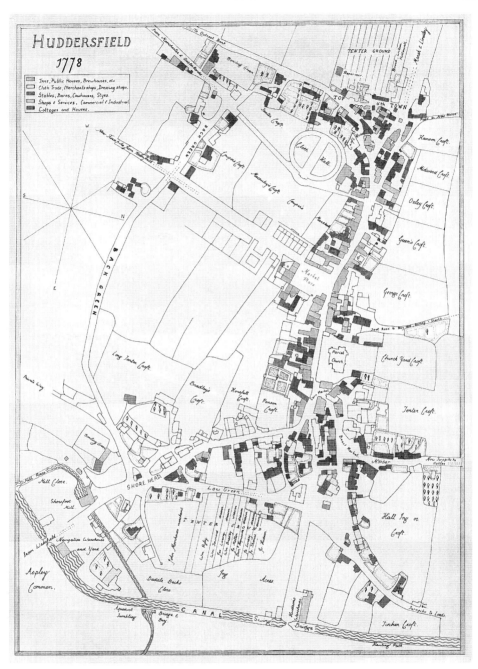

This plan of Huddersfield, redrawn by GR from the Ramsdens' estate map of 1778, illustrates the growth that had taken place since 1716 at both ends of the town street. It is most noticeable at the top of the town – today's Westgate. Information has been included from the survey to demonstrate the functions of the buildings and the extent of the town's commercial development, a point emphasised by the tenter grounds, Cloth Hall and canal.

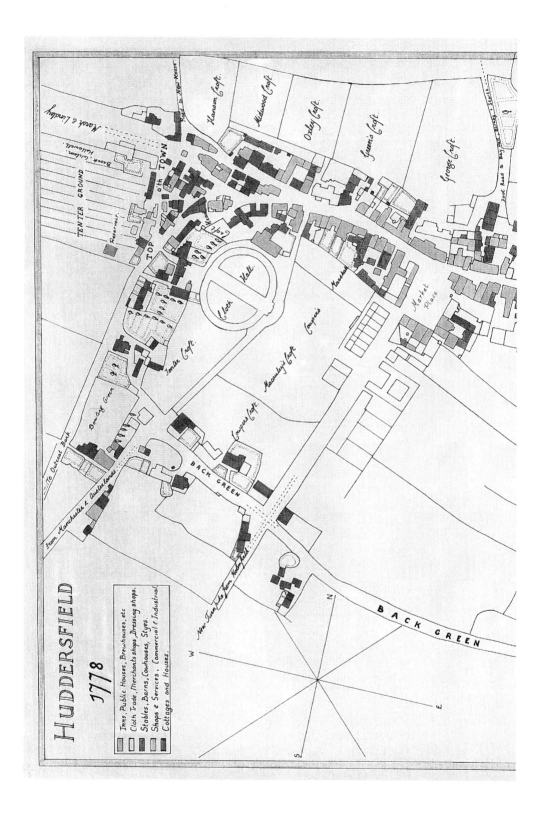

HUDDERSFIELD
1778

Inns, Public Houses, Brewhouses, etc

Cloth Trade, Merchants shops, Dressing shops.

Stables, Barns, Cowhouses, Styes.

Shops & Services, Commercial & Industrial

Cottages and Houses.

TENTER GROUND

Brook, Colhorn Holmfirth

Marsh & Lindley

Road to New House

Hanson Croft.

Midwood Croft.

Oxley Croft.

Queen's Croft.

George Croft.

Road to Bay Hall, Leeds, Sheffield

'Oth TOWN

TOP

Reservoir.

Cloth Hall

Marsh

Market Place

Tenter Croft.

Bowling Green

TOWN

To Outcast Bank

From Manchester & Austerlands

Macaulay's Croft.

Cooper's Croft.

Cooper's

BACK GREEN

New Turnpike from Holmfield

BACK GREEN

N

W

E

S

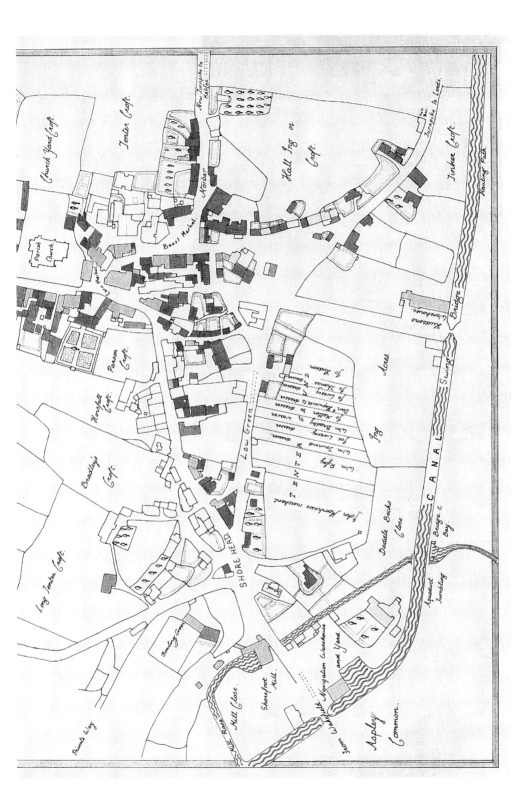

LUCK LANE (Marsh)

There are several references to Luck Lane in the 18th century, and sources include title deeds, the Ramsden estate surveys of 1716 and 1780, and the Enclosure Award of 1789. I think the name should be seen in conjunction with Luckeroyd and Luke Bank, two minor names recorded in that area in 1716. These were both at Paddock Head, at one end of Luck Lane, and they point to the personal name 'Luke' as a possible source. The suffix 'royd' suggests an assart or clearance dating back to the 13th century or earlier, so 'Luke' may have been the tenant responsible. There is a reference in the court roll of 1532 to a place called 'Lukoxgang', and this is likely to be the same man.

MACAULEY STREET

The Macauley family seems likely to have arrived in the Huddersfield area soon after 1700. Alexander Macauley had a son Archibald baptised in Elland in 1707 and there are deeds to show that he was the owner of a new house in Huddersfield Market Place in 1711. He was the direct descendant of Sir Auley Macauley, the laird of Ardencaple in Dumbartonshire, but he established himself in the town as a mercer or linen-draper and had several sons. George Gibson Macauley, the Yorkshire cricketer, was a direct descendant. There is a reference to Macauleys Buildings in a directory for 1809-11, and another to Macauley Street in 1821. Stephen Macauley still resided there in the 1850s. Part of the street has survived a succession of town centre improvements and it marks where the family had cloth-dressing shops in the 1790s.

MANCHESTER STREET

This was the name of a street that formerly ran from the top of High Street towards Outcote Bank, but no trace of it survives and the land is now occupied by the Civic Centre. It was in existence before major alterations to the Manchester turnpike road, c.1820, and no doubt owed its name to the fact that it was part of the direct route into Lancashire from the top of the town. It is called the Road to Manchester on the town map of c.1780. In 1808 Joseph Heywood was granted a lease on new property between Upperhead Row and Manchester Street, but only on strict conditions. In view of the way in which the town was growing in this period it is worth quoting these in detail. He was not to 'exercise the trade of a blacksmith, farrier, tanner, skinner or chymist' and his house was not to be used 'for a steam or fire engine … as a slaughter house, or a place for making pots or tobacco pipes, burning of blood, making of glue or sizing, soap or candles'. Nor was it to be used as a place to sell cloth 'or any flesh or butchers meat on any market day'.

MARKET AVENUE or GREENWOOD'S YARD (New Street)

Greenwood was a well-established surname in Huddersfield in the early 1800s but I cannot find a direct link with Greenwood's Yard which ran between New Street and Victoria Lane. It is shown on the OS map of 1851 and can be found in directories from 1853 but the name was changed to Market Avenue some time before 1934. More recently this has become a covered shopping centre but there is still evidence in an upper storey of a warehouse building with its hoist attached.

MARKET PLACE, MARKET WALK

Huddersfield acquired its status as a market town in 1671 and it is likely that the Market Place was created almost immediately. In 1677, for example, when money was being raised in the town towards the rebuilding of the king's 'ships of warr', Richard Williamson's assessment included 1s 6d 'for benefit of Markett Place'. This was an open market, operated in accordance with numerous local traditions, and its early history has been well told by Edward Law. Extracts from the wills of Richard Williamson and his widow show that they were the 'farmers' of the market, and an inventory of 1696 lists 'the Brass Stroake, the yardwand and the Cryer Bell', all items necessary for its administration. We know from title deeds and the survey of 1716 that much new property had by then been built around the Market Place, and this reflected its commercial success, so important to the town and the landlord. We can picture it on a busy market day, for a lease of 1718 mentions 'all the Stalles and Trustles' that belonged to it, and numerous Quarter Sessions documents paint a vivid picture of the activity there. It is uncertain just how old the narrow alley is that links it with King Street but maps show that it has been there from the early 1800s and the official name Market Walk is recorded from 1826. Unofficially this was known as 'Wappy Nick'.

MARKET STREET

This street was built in the last quarter of the 18th century, probably in connection with the new Cloth Hall, and it was initially given the name Cloth Hall Road. However, it had become Market Street by the end of the century, even though it had no direct connection with the Market Place. It may simply be that the Ramsdens thought of it as a suitable name for the growing town and gave it their approval because it drew attention to their new 'market' - the Cloth Hall.

MARSH

In a deed of 1436 Marsh is referred to as an area of waste and common. It lay to the west and north-west of Huddersfield and was probably a fairly barren area, wet and boggy if the name is to be trusted. It probably did not include the medieval settlements of Edgerton and Gledholt which are thought to have had 'hamlet' status in the township of Huddersfield at that time, that is to say they enjoyed a degree of autonomy. In fact there is no evidence of settlement in Marsh until the second half of the 16th century when important enclosures took place and several cottages were built. The earliest residents included families called Batley, Dyson, Gledhill, Hirst, Mallinson, Smith and Sykes. By the mid-1600s these scattered dwellings had been given distinctive names, such as 'Coit', 'Lathe', 'Lane end' and 'Cross', but many of these did not survive for very long. Two names that did survive were Blacker Lane and Heaton Fold, and they remind us of two other families who made their homes in Marsh in that early period.

Marsh started to be thought of as a territory in its own right in the 1600s, independent in some respects of Huddersfield. The evidence for that occurs in disputes about the rates and the maintenance of its highways, first in a manorial roll of 1657 and later in Quarter Sessions documents. In 1692, for example, Alice Hirst of Gledholt complained that she was being assessed as though she belonged 'to the toune of Huddersfield', although she was an 'Inhabitant in a nother hamblit … called Marsh'. Her petition also confirms that Gledholt had by then been absorbed into the 'hamlet' of Marsh, as had Paddock. Although piecemeal enclosure of Marsh Common continued through the 18th and 19th centuries, with new intakes recorded in successive rentals, the territory retained something of its rural character into the 1900s. Now, however, it is almost completely urbanised.

MARSHALL ROW

The only evidence for a street called Marshall Row is on the map of c.1780, roughly where Half Moon Street now is. Joseph Marshall held a block of property there in 1778, on the south side of the street, so it is quite likely that Marshall Row was an unofficial name, directly comparable with Midwood Row. Its history should be seen as an extension of that for Croft Head. See Snow Hill.

MARSH CROSS

All trace of this cross seems to have disappeared, along with the place-name. It is shown on Jefferys' map of 1772 and the Ramsden estate maps of 1716 and 1780, and can be seen to have been located close to where Eldon Road now joins Westbourne Road. In fact, the older name for Eldon Road and Dingle Road was Cross Lane which linked Heaton Fold with the old highway to Lindley. The cross probably marked this junction and is unlikely to have had any religious significance. In 1716, Cross Lane was still a track over rough common, although there were one or two intakes and buildings to the west and Mr Wilkinson's freehold to the east. A Marsh family called Hirst was said to be 'of Cross' in the 1600s, and the cross itself is referred to in a deed of 1519.

MARSH GROVE ROAD

The suburban streets that now link Cleveland Road and Westbourne Road are called Marsh Grove Road and Croft House Lane, and where these two come

together marks the approximate site of two former villa residences called Marsh Grove and Croft House. These are clearly shown on the OS map of 1854, when John Freeman, gentleman, and Thomas Brooke, solicitor, were living there. At that time the two properties were surrounded by fields, but all that changed after 1887 when the Croft House estate was put up for sale. Marsh House is shown on the OS map of 1843 and it survives as the name of a public house on Westbourne Road.

MASONS' ARMS
Little is known about this public house which formerly stood in the Corn Market. It was called the *Maysons Arms* in the Quarter Sessions in 1777, and the estate map of 1778 shows it next to the *Blackamoor:* John Kaye was the landlord at that time. According to Edward Law it was demolished in 1802.

MAZE (Westgate)
On the map of c.1780 'the Maze' was at the top of the town, at the junction of routes to Gledholt, Lindley and Outcote Bank. It was presumably a reference to the network of lanes and alleys there, which were swept away in the early 1800s. The site was roughly where the railway line now passes under Westgate.

MIDWOOD ROW (Westgate)
The only evidence for this name is on the map of c.1780, in the location later known as Temple Street. The Midwood family had a long history in that part of the town and James Midwood's tenancy there is detailed in the survey of 1778. The estate map shows that he held most of the buildings on the north side of the street, including his own residence and various shops, warehouses and work places. The place-name may have been used informally.

MILL GATE (Paddock)
I have no early evidence for this name which refers to a lane at Paddock Foot, close to the river Colne. However, there have been mills on the site for at least 500 years, starting with the fulling mill run by the Thornton family, so the name Mill Gate may have been in oral use throughout that period.

MILN HILL
This was the name of the hill on which the school at Seed Hill was built. The evidence for it goes back to 1681, when Sir John Ramsden of Byram leased a 'Parcel of Ground called the Milne hill … near Huddersfield Town Mill' to a group of four trustees, as the site for the new school. There are earlier references to 'Miln Hill' locally but the place-name is frequent in Yorkshire and most of those that I have come across seem to apply to Mill Hill in Dalton.

MOLD GREEN (Dalton)
This part of Dalton, first referred to in 1519, became more closely associated with Huddersfield after 1850, when land there was put up for sale by the trustees of the Lister Kaye estate. The trustees' intent is implicit in the description of Mold Green as 'well situated for the convenience of Trade … well roaded, and

Old Bank Fold; an attractively renovated part of Mold Green.
The 'bank' was part of the former highway to Wakefield, en
route to Almondbury.   G Redmonds

... within a short distance of the improving town'. The momentum quickened after 1868 with its inclusion in the borough and by the end of the century it had become a suburb of Huddersfield. Initially, it had been an area of common grazing, as the suffix 'green' suggests, and early references are to the illegal digging of stone (1581) and feeding of geese (1655). The prefix 'mold' may point to the practice of extracting alluvial black earth from the areas around Carr Pit.

MOLDWARP HALL

Although this house is located close to the Bradford Road, at the lower end of Lightridge Road, it is easy to imagine it in its pre-turnpike rural setting. The first reference that I have located to the name is in 1771, when it was called Mouldwarp Hall and Henry Hey was the tenant. 'Moldwarp' was the dialect word for the common mole, a creature that threw up piles of earth overnight, so the name may be a jocular reference to an early and unauthorised encroachment on this part of the common – a similar name therefore to Mushroom Hall. Later, perhaps because the tenants were embarrassed by the name, it was called Mole Hill but the present owners happily use the original name. On the OS map of 1854 there was a ford close by, on Judy Lane, but I cannot link this name to any particular person.

MOUNT (Birkby)

See Ark Hill Mound.

MOUNTJOY ROAD

This street is on the line of a much more ancient right of way than we might at first think. For example, it is called the Old Lane on the OS map of 1854 and it ran from Snodley in Marsh to Highfields and New House. As earlier maps show, it predates the turnpike road that we call New Hey Road and was actually part of the ancient highway between the top of the town and Lindley. If the road commemorated an individual or family called Mountjoy, the ultimate origin would be a French place-name but it probably derives from Mountjoy House, the villa home of Mr William Fairweather in 1879. If that is the case it may have a similar meaning to Mount Pleasant, a name discussed by Adrian Room in The Street Names of England (1992). This was sometimes used ironically but also suggested 'an elevated and salubrious locality', suitable for the residence of a gentleman.

NELSON'S BUILDINGS (New Street)

The term 'buildings' is used in numerous minor names in and around the town and most of them date from the 19th century. Often they commemorate an enterprising individual and that was true in the case of Nelson's Buildings remembered by local historians because they were the Headquarters of the Mechanics' Institution from 1844 to 1850. They were on the west side of New Street, between Hawksby Court and the bottom of Cloth Hall Street. Mr Schofield's reminiscences of the period 1825-26 really evoke this bit of old Huddersfield: 'Henry Nelson, Esq., lived in the building now bearing his name, and kept a good staff of servants and assistants. The front of the house was railed off, and it was covered with sweet jessamine and other creepers'.

NETHER BRADLEY

This was an alternative name for Bradley, used quite commonly in the 16th and 17th centuries. Professor Smith suggested that it served 'to distinguish it from Bradley Mills' higher up the valley but that is not the case. It actually distinguished it from Bradley in Stainland, sometimes called Over Bradley. Henry Savile was said to be 'of Over Bradley' in 1577 and I suspect that the prefixes were useful because the Saviles had interests in both places.

NETHER GREEN

In the 16th and 17th centuries this was the name of the common or waste at the lower end of the town gate, below what is now the Beast Market. The earliest reference is to 'the Nethirgryne of Huddisfild' in 1510, and the name occurs several times after that in connection with encroachments and enclosure. In c.1570, for example, Robert Woodhead was granted 'a third parte of an acre inclosed from the neither grene', and an undated deed of the same period mentions 'a messuage from waste land in Nethergreyn ... newly constructed by William Batley'. A poor man called Joseph Brooke was said to be 'of Nethergreen' in the hearth tax returns of 1672 but I have no later examples of the name and it gave way soon afterwards to Low or Lower Green.

NETHER HOUSE (Birkby)

This name has apparently not survived. It referred to a dwelling next to the corn mill that was once part of the Clough House estate, in the possession of Elizabeth Brooke in the 1620s. The name describes its location in relation to Clough House itself and it had actually changed to Lower House by the 1700s. Later, textile mills and then a supermarket were built on the site.

NETHEROYD HILL

The modern spelling is deceptive, for the first element is not the word 'nether'. The early form Nase Rode (1532) suggests that it may derive from the Old English word 'nasu' meaning nose, in which case it was probably a 'royd' on a promontory. Typical 17th century spellings, such as Nathroide and Naithroyd, gave way in the 1700s to Nethroyd and eventually Netheroyd, the present form. This final development is likely to have been influenced by the surname Netherwood, for a family with that name settled in the area in the late 1600s. In 1861, for example, Henry Neatheroyd was living at 'Neatheroyd Hill'. The area was part of the common that extended from Sheepridge to Cowcliffe, and the earliest reference to it records an enclosure made by the Brook family in 1532. The first evidence of a settlement there is early in the 1600s when families called Hirst and Brook were described as being 'of Nathroide'. This suggests that cottages had been built on the encroachments. It was a development that continued right up to the Industrial Revolution. For example, in 1706, the magistrates granted 'Elizabeth Armitage, widow, liberty to Erect a Cottage upon the Wast ... near Nathroyd' and the rental of 1716 records that widow

Sykes had cottages there. The landscape, together with place-names such as Stone Delves, testify to the fact that stone was quarried in this area over many centuries, but documentary evidence for that is scarce. We have to wait until the late 1700s, when the accounts of the Fartown surveyors of the highways include entries such as 'filling a delf on Netheroyd Hill' (1789).

NEW DROP (Cowcliffe)
Smith listed this among the township's field names but on the OS map of 1854 it clearly refers to a building. That is less clear on the earlier 1 inch map (1843). The name is found elsewhere in the West Riding, so it may have been a grim reference to the gallows. 'New Drop' is quoted with that meaning in the OED from 1796.

NEWHOUSE HALL (Sheepridge)
This house owes its name to the fact that Thomas Brooke of Deighton was granted a piece of land on Sheepridge common in 1521. This was described as 'lying to the west of a gate called Bradley Yatte, to the south of Bradley Wood, to the north and east of the said common'. In his will of 1554 Thomas referred to the land and to the 'new house ... lately builded' and the inference is that it had been put up soon after 1521. However, the earliest record that I have of the place-name is 1533, when Thomas Brooke 'of Newhouse' was named in a list of local clothiers. His descendants remained at Newhouse for several generations, adding to their property as the occasion offered. For example, a lease of 1611 took account of 'the expenses of Thomas Brooke ... in taking in, enclosing, fencing and reducing into husbandry one piece of barren ground of the waste ... called Sheepridge'. Between 1590 and 1630 there are references to the 'nethernewhouse' and the 'overnewhouse', and these may at first seem to suggest that a second house had been built on the site. However, the named tenants in that period were called Brook and there are good grounds for believing that Newhouse may simply have been partitioned c.1590, with the old timber building incorporated into a larger stone structure. In such a case 'over' and 'nether' would simply refer to the two ends of the building. By 1630 the family had acquired gentry status but there is no evidence that 'hall' was added to the place-name in recognition of that fact. Indeed it seems more likely to date to the 1860s when Sir John William Ramsden purchased the property and had the east wing rebuilt.

NEW HOUSE (Highfields)
The history of this house has already been well told by Edward Law. It lay within a short distance of the town centre, near Highfields, on a site that commanded 'a view of the town' with 'a most beautiful and extensive prospect over the surrounding country'. It was the home of a family called Bradley, and the suggestion is that it was built c.1720, on land that the family had owned for many years. The earliest reference to the place-name that has been noted occurs in the parish register for January 1727. There are still interesting listed buildings at New House and the place-name survives in Newhouse Place.

Newhouse Place; part of the high-quality development at Highfields. It owes its name to nearby New House, built in the early 1700s by the Bradley family. G Redmonds

NEW NORTH PARADE

This name has been given to the surviving section of the former Halifax Road which branches off Westgate by St Patrick's Club and ends at the Ring Road. It seems to be a compromise between New North Road, of which it was originally a part, and North Parade. The latter name goes back to 1837 at least when it is listed as part of the New North Road in White's Directory. Unfortunately North Parade's earliest history and status are far from clear, since the name is missing from many maps and directories. It seems possible that it became an unofficial 'prestige' name when the new Halifax Road was named New North Road.

NEW NORTH ROAD

This important thoroughfare was constructed in the early 1800s. The line of the road is shown on the map of 1820 and it had been completed by 1826 when it was named on Crosland's map as Halifax New Road. Almost immediately a number of villa residences were built there, taking advantage of the elevation and the proximity to the town centre. This established its character which was further emphasised when development began at Edgerton. The name New North Road is found from 1837, and White's Directory of 1853 described it as 'a very handsome entrance to the town' with 'a great variety of architectural designs'. The trustees of Lewis Fenton (1857) were confident that their development of 'first class villas' there would be successful, since it was 'the most attractive (locality) in the area', much preferred by the public. Murray Road was part of this scheme.

NEW STREET

There were several significant developments in Huddersfield in the period 1759-1780, notably the new turnpike roads, the Broad Canal and the Cloth Hall. Even at the time it must have seemed a pivotal period in the town's history, part of its transformation from a busy village on the fringe of the moors to a major commercial and industrial centre. The building of New Street should be seen in this context for it created a new axis north to south, a complete change from the ancient alignment of the old town gate. Moreover it was broad and straight, another feature of its 'newness'. It probably began as the first leg of the new road to Woodhead in 1768, starting at the Market Place and heading due south towards Chapel Hill and Engine Bridge. There may have been an earlier footpath across the fields in that direction but the first real evidence of a road is on the town maps of 1778-80. The fact that it was called a 'street' is significant for we know that this word entered the local vocabulary in the 1650s, with the first references to the 'town street'. On the west side of the street, from the Market Place to the bottom of Cloth Hall Street, was a range of buildings known originally as the New or Brick Buildings, prestigious town houses in their day. It is said that they were constructed out of bricks that were not needed for the Cloth Hall, although today the New Street façade is covered in painted stucco. Not all the houses survived: those at the south end of the range were demolished as early as c.1880, to make way for a bank that survived until 1971. The name New Street was in use in rentals from 1798 and it is also found on the map of 1797, although I believe it was written there retrospectively. The original intention may have been to call it South Street or Southgate, for both these names feature on the early estate maps.

NEW TOWN, NEWTOWN ROW

Alexander Hathorne was an agent for the Ramsden family and in 1849 he wrote to his superiors about a plan to lay out what he called 'the new town of Huddersfield'. He was discussing the role to be played by St George's Square, part of the development that would result from the building of the railway station and the demolition of the George Inn. That phase of 'Victorian' improvement is still obvious, even to a casual observer, if only because of St George's Square and the regular layout of the streets to the north of Kirkgate and Westgate. Not surprisingly, writers on Huddersfield tend to talk of the 'New Town' when they discuss this development, and yet the name has a much longer history. On the map of 1826 there is actually a grid plan of streets described as 'intended' or 'not made' roads, most of them to the east and west of Northgate. It is difficult to say what happened to that project although Belmont Street appears to preserve the west end of the 'not made' road. The name is found on a tombstone in the churchyard dated 1809 and again in the trade directory of 1822, where Godfrey Berry was said to be 'of Newtown'. Newtown Row seems to have been the later name of the street listed as Newtown in early directories, and it is shown in 1851 as a terrace of eight properties, each with its own yard. It lay to the south of Oxford Street and its western end would have been close to Clare Hill.

NOOK
A family called Horsfall lived at 'Nook' in the 16th and 17th centuries, but I have been unable to find out where it was. There were other Horsfalls at Copley Stones, so the farm may have been in that part of Fartown.

NORBAR
This is one of a very few old town centre place-names, and it was first recorded as Northberre in 1573. Other early spellings include Norbarr (1585) and North Barr (1599). The map of 1716 shows that these referred to a location beyond a row of houses in the area that later became Northgate. There was a house at Norbar in this period, possibly the 'mansion house' occupied by William Taylor in 1606. However, the name would have referred originally to a barrier of some kind, one that marked a division between dwellings to the north of the town and ancient fields such as Sally Carr and Kirkmoor. The 'bar' is likely to go back well before 1573, possibly to the time when the open fields were created. Interestingly, an identical place-name occurred in other Yorkshire towns, in Leeds for example which also had its East Bar. In this connection it is worth noting the by-name John atte Barre of Wakefield in 1362. Having first described a barrier and then a dwelling house, Norbar came to be associated later with the lane that extended northwards from the bottom of the Beast Market, and in the court rolls we find references to Norber Lane (1731) and Norbery Lane (1715). The latter spelling suggests that by this time Huddersfield people no longer had any perception of the name's original meaning. Nevertheless it remained in use into the 1800s, even after it was replaced by Northgate.

NORMAN PARK, NORMAN ROAD
The Council Proceedings show that it was resolved on 4 August 1896 that 'Fartown Recreation Ground be named Norman Park', taking its name from Norman Road near by. Over two years earlier, in April 1894, the Town Clerk had written to Col. Beadon, who was then the Ramsdens' local agent, to inform him that the Corporation had previously purchased land from Mr Scholes in that part of the town and had plans to lay it out 'for recreation'. Sir John Ramsden agreed to contribute £200 towards the cost of its development, and over the next two years the stream was diverted, partly covered over, and a lake constructed. Embankments were built and land 'below the terraces' was levelled and laid down in grass. Not everybody was pleased. A certain Mr North had wanted to use the lower portion of the ground as a poultry run but was informed by the Borough Surveyor that his application could not 'be entertained'. In 1905

Norman Park was to be at the centre of a long trial in the High Court of Justice. A violent storm the year previously, on 24 July 1904, had been responsible for damage done to property in many parts of the town but Middlemost Brothers claimed that disastrous flooding at their Clough House Mills was the result of work carried out in Norman Park on behalf of the Corporation. The depositions are of real interest for the references they make to the park and to other places in that vicinity, such as Storth, Jack Bridge, Hebble Beck, and Allison Dyke.

NORTH CROSS ROAD

This road runs westwards from Cowcliffe Hill Road, just to the north of South Cross Road: they both take their name from a cross that once stood on Cowcliffe Common and marked an important junction of highways. The base of the cross survives, partly hidden in the undergrowth, and it is the only visible evidence we have of several similar crosses in the township. The road name seems to have its origins in the arrangements made during the Huddersfield Enclosure of 1789: it was an 'occupation' road and lead westwards over the north side of the common to certain tenements and an ancient field called Upper Intack. This former intake is shown on the map of 1797, roughly where Cowcliffe Plantation now is.

NORTHGATE

Northgate was probably so named late in the 1700s, having replaced 'North Street' which is on the town map of 1778. It is shown there as part of the new and more direct highway to Elland and Halifax, a route that incorporated the former Norbar or Norbery Lane. It is difficult now to picture the lane as it was into the 1700s, with its gardens, orchards and hedgerows. It remained an attractive place to live even after it became part of Northgate, for the mansion called Carr House was situated on its western side. Much of Northgate disappeared when the Ring Road was built but at least the place-name has survived, restricted to the stretch between Lower Fitzwilliam Street and Bradford Road

NORTH PARADE

See New North Parade.

NORTHUMBERLAND STREET

If we are to make sense of the earliest references to Northumberland Street we need to know that it was constructed over a period of ten or so years starting in the 1840s, and that it was formerly much longer than it is at present. It now ends at the Ring Road but originally it continued east beyond Northgate, stretching as far as the Old Leeds Road. That site is now occupied by Ibbotson Flats. In fact the first section of the street to be completed ran from the Old Leeds Road to the intersection with Primitive Street, a name that commemorates the Primitive Methodist Chapel of 1846. As early as 1849 the records of the Paving Committee refer to numerous owners and occupiers of property in Northumberland Street, but these were residents in the section that has now disappeared. References from 1850 to an 'intended new street' called Northumberland Street had to do with the extension westwards towards the new George Hotel. Work on levelling this way through the fields, followed by the installation of drains and paving, was taking place through the 1850s. Now, we associate Northumberland Street with the Post Office, a fine building that was opened in 1914 and dominates the south side of the street, almost directly opposite its predecessor. I cannot yet say whether 'Northumberland' is a reference to the county or to the title, but construction on the street began in the 1840s when the Trustees were running the estate, so one of them, or one of the Ramsdens' local agents, may have influenced the choice of name.

OAK, OAK ROAD (Bradley)
Oak Road in Bradley may be a relatively recent name but it reminds us of one of the oldest farms in that part of Huddersfield. On 19th century maps Oak Lane was a right of way that branched off the loop of the old highway from Leeds and made its way towards the woodlands further to the west. Their names in 1829 were Lodge Wood and Oak Wood but Lodge Wood was felled some time before 1854 and Oak Wood had been renamed Dyson Wood, presumably after the tenant William Dyson. Since then further clearance of woodland has taken place and what survives of the lane is known as Oak Road. If we go back even further, into 17th century records, we find a reference to Oak Lane in the Quarter Sessions (1691) and another to a Bradley farm called 'Oke' in the court roll of 1630. At that time Thomas Hirst was the tenant and he was followed by John Hirst 'of the Oake in Bradley' who made his will in 1680. A lovely old house, dated 1751, survives at the junction of the former lane and the highway, so this may be the site of the original farm. However, it is not named on the map of 1829 when Francis Spence was the tenant, although two cottages further to the west were called Oaks.

OAK WOOD (Bradley)
In 1704, Sir Lionel Pilkington sold the spring woods on his Bradley estate to the owners of the iron forge at Colne Bridge. The woods were all listed individually, that is: Lad Greave, Oakwood, Hirst Spring, Pellet Royd, Bradley Yate Wood, Newhouse Wood, Park, Ambler Wood, Overworth Wood and Fell Greave Wood. Now, much of the woodland and many of the names have disappeared and some of that had to do with the closure of the iron forge and the building of the canal and railway. Later, intensive housing development was largely responsible.

OCHRE HOLE (Fixby)
This is another Fixby place-name with strong Huddersfield associations. Ochre was used as a dyestuff by clothiers in this area from the 17th century at least, and there is evidence that it was mined locally. As the place-name is referred to in a court roll as early as 1688, it may be evidence of an early small-scale industry in that area. In later examples the name was linked with 'lane' and 'clough', and a Quarter Sessions document of 1739 describes 'Hocker Hole Lane' as part of the highway leading between Huddersfield and Halifax. The name is still generally known as Ocker 'oil, a pronunciation that keeps faith with the dialect. However, the initial 'H' had also been used in 1688 and the feeling is that it represents a determination by the clerks not to use the dialect.

OLD BANK CHAPEL

The Wesleyan Methodists of Huddersfield acquired a site on which to put a chapel in 1775, and it was built the following year, of brick. It is shown on the town map of 1778 as 'Dissenters' Chapel', away from the centre of the town on the hill or 'bank' that led to Lockwood. It was the only Methodist chapel in the central area until 1797 and it was also known as Buxton Road Chapel. It gave its name to Chapel Hill and was rebuilt in 1837, during a period of growth and prosperity. It was partly because of its early foundation that it later came to be known affectionately as the Old Bank Chapel. Its doors closed in 1950.

OLDGATE, OLD STREET

Formerly, the town gate turned almost abruptly south just below the church, on its way to the corn mill at Shorefoot and the bridge across the river Colne. Now, the latter part of this ancient route has been absorbed into the Kingsgate development, and traffic leaving the town in that direction is obliged to follow the Ring Road. A remnant of the street survives as Oldgate and yet this is not a very old name. On the town map of 1778 it is called Bridge and Mill Street, almost certainly a name that was coined at the time of the survey, since it is found in no other document. Part of it may have been the 'cawsey' referred to in 1642, but in the 19th century it was more commonly Kirkgate, Old Kirkgate or just Old Street.

OLD POST OFFICE YARD (Kirkgate)

See Post Office Yard.

OLD SOUTH STREET

See South Street.

OUTCOTE BANK

In the Huddersfield area 'bank' is the usual word for a steep slope, especially where there is a right of way. Outcote Bank was part of the highway from Huddersfield to Manchester and it is recorded in the Quarter Sessions as 'Outcoate Banck' in 1656. It owed its prefix to a 'cote' built on the waste, 'out' of or away from the town itself. This would originally have been either a small dwelling house or an animal shelter, although the name was actually that of an enclosure in a title deed of 1545. Much of the bank was common land, and estate rentals show that settlement and enclosure were still taking place there in the 18th century. In 1716, for example, there were payments of 5s for 'a poor Cott. on outcoat banck' and 10s for an 'Intack on outcoat banck'. It remained a relatively undeveloped area into the early 19th century, and the parish register records how William Bland was 'found dead in the Outcoat Bank' on June 3 1811.

OUTSHOT

Although this is not really a place-name, the term is of interest on two counts, firstly because it concerned a town centre property and secondly because it was linked by Philip Ahier with Outcote Bank, incorrectly I believe. In Yorkshire an outshut or outshot was an extension to a building, and examples of the word are

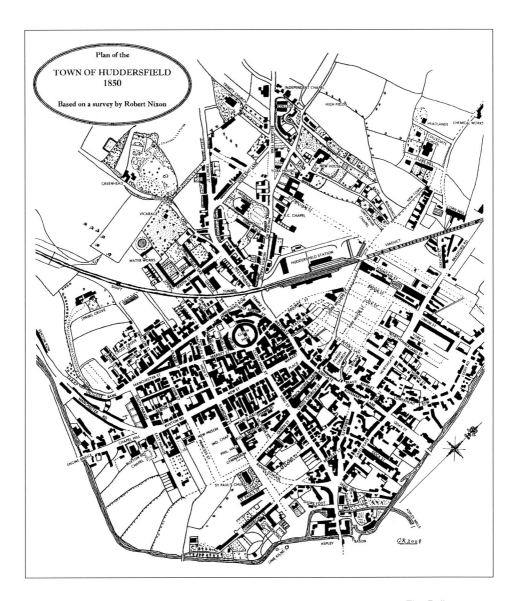

Huddersfield in 1850; a map redrawn by GR from Robert Nixon's survey. The Railway Station is a prominent feature and work is beginning on the New Town, but the street plan illustrates the development since 1820 under the Commissioners. Schools, churches and chapels reflect the growth in population, whilst the expansion along Trinity Street and the new Halifax Road is clearly visible. Nevertheless green fields survive, as do the important estates at Greenhead and Spring Grove.

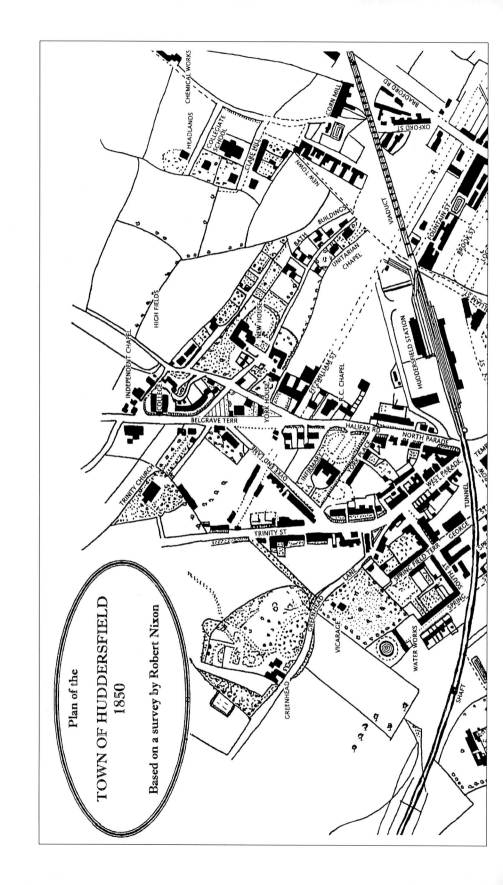

Plan of the

TOWN OF HUDDERSFIELD
1850

Based on a survey by Robert Nixon

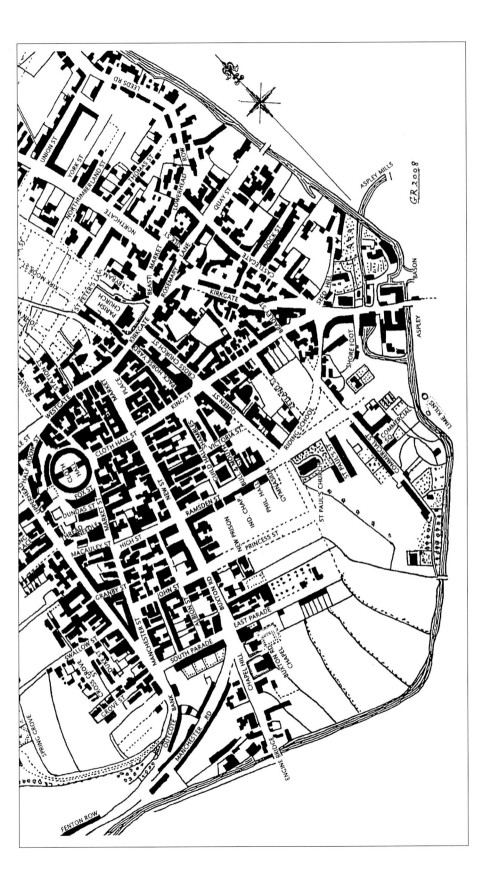

GR 2008

FENTON ROW
SPRING GROVE
OLLCOTE BANK
MANCHESTER RD
BUXTON ROAD
CHAPEL HILL
ENGINE BRIDGE
SOUTH PARADE
EAST PARADE
CHAPEL
GROVE ST
CROSS GROVE ST
SWALLOW ST
MANCHESTER ST
JOHN ST
ALBION ST
GRANBY ST
HIGH ST
MACAULEY ST
NEEDLE
HEAD
DUNDAS ST
FOX ST
CLOTH HALL
NEW ST
MARKET ST
CHILD ST
RAMSDEN ST
PRINCESS ST
NEW PRISON
IND. CHAP.
PHIL. HALL
GYMNASIUM
BUCK ST
VICTORIA ST
QUEEN ST
KING ST
CLOTH HALL STREET
RIDING SCHOOL
ST PAUL'S CHURCH
ST PAUL'S ST
COMMERCIAL ST
COMMERCIAL SQ
LIME KILNS
BASON
ASPLEY
SHORE FOOT
SEED HILL
BLADE ST
BEARD'S ST
CASTLEGATE
DOCK ST
KIRKGATE
CROSS CHURCH YARD
PACK HORSE YARD
MARKET PLACE
PARISH CHURCH
ST PETER'S LANE
KIRKGATE
BYRAM ST
BEAST MARKET
ROSEMARY LANE
LOWERHEAD ROW
THOMAS ST
QUAY ST
MARKET ST
NORTHGATE
NORTHUMBERLAND ST
YORK ST
UNION ST
LEEDS RD
KIRK MOOR ST
JOHN ST
RAILWAY ST
WESTGATE
HALF MOOR ST
HEAD
BREAD
KIRKGATE ST
HALF MOON ST
ASPLEY MILLS

recorded locally from the early 1500s. A deed of 1543 mentions an 'owteshott' in Huddersfield that lay near the cemetery – presumably the same building that was described in 1589 as adjoining the cemetery and 'buttinge upon the towne gate'. It was then in the occupation of the Blackburn family. It may have been 'the Hous and Pentice over the Church Yard' listed in the rental of 1716.

OVER GREEN
This was the green or waste at the western or top end of the town gate, first referred to in the enclosures of c.1570. Widow Blackburn was one of several tenants who acquired land there at that time, taking possession of 'half an Acre by estimacion inclosed from the over gren'. Later it became the Upper Green, and this name was still in use at the end of the 18th century. In 1789, for example, there was a carriage road from 'the Top of the Town ... over the Upper Green to an ancient lane leading to Dyke End'.

OVERWORTH WOOD (Bradley)
There is a single reference to this name, in a document of 1704, so I can only speculate about its meaning. I believe that it may originally have been Overthwart Wood, descriptive of the wood's location in relation to its neighbours. 'Overthwart' meant 'across' but was regularly spelt 'overwhart' in this area and the transition would have been helped by the popularity of 'worth' as a suffix.

OXFORD STREET
This was always a short street, and the map of 1850 shows it running between Fitzwilliam Street and the railway viaduct. It was part of the New Town development, linked with Green Street and Newtown Row, and must still have been in the planning stages in 1850. In fact, the Commissioners' records show that the levels of these streets were still being fixed in November 1852, in preparation for the pavements. The choice was probably inspired by Oxford Street in London, a name which may not be as straightforward as it seems. There was a highway there in the 1600s that was actually part of the road to Oxford but it became a street in the 1720s when it was part of the estate of the second Earl of Oxford. The name was popular in the northern industrial towns, anxious as they were to demonstrate their growing importance, and the Trustees were no doubt attracted by its aristocratic associations.

PACK HORSE (Kirkgate)
The original *Pack Horse Inn* was one of the town's most important and best-loved hostelries, especially the 'Mews' in the Pack Horse Yard, but despite that the buildings were demolished to make way for the Pack Horse Centre. The meaning of the name is transparent and the records show that the inn was at the heart of Huddersfield's commercial development. D.F.E. Sykes described the inn yard as many local people can remember it – 'approached by a high gateway, more than two hundred yards in depth'. At the height of its prosperity there were stalls for a hundred horses, granaries for the corn, accommodation for stage coaches, post-chaises and carriers' waggons.

PADDOCK

Territorially, Paddock has been part of Marsh throughout its history, and the fact that it means 'small enclosure' suggests that it was originally a clearance in that part of the common. Characteristic spellings, as late as the 18ᵗʰ century, were 'Parocke' (1583) and 'Parrock' (1771) but examples of the modern form occur towards the end of that period, e.g. 'Parrockfoot alias Paddock Foot' (1766). The first references to the place-name point to a very small and scattered community, one that already had a 'top' and a 'bottom'. In 1568, for example, Edmond Hyrst was said to be 'of Parockfote' and just a few years later John Battley enclosed land 'on the head of the parock so called'. It remained a scattered community for centuries, as the manorial survey of 1716 demonstrates. In that year tenants had houses at the East End, the Upper End and the North Side. Mr John Hodgson's cottage was 'at the Nabend of the Parrack', a place-name that has survived. Down by the river was a fulling mill, called just the 'mylne' in 1598 but more explicitly 'Parockfoote' mill in 1625. The Thorntons were the family most closely associated with it throughout its early history.

PEEL STREET

This is a short street running between Ramsden Street and Princess Street, immediately to the east of the Town Hall. It was completed in 1881 but work had begun three years earlier and Peel Street is listed in the directory of 1879. It was there that the Borough Police Force had its headquarters, and the source of the name seems certain to have been Sir Robert Peel. He was the statesman responsible for the London Police Force when he was Home Secretary. A statue of him was unveiled in St George's Square in 1873 but the stone became badly corroded and it was removed in 1949.

PELLET ROYD WOOD (Bradley)

This former name for Screamer Wood is one of the earliest documented names in the township. A Fountains Abbey charter of the 1190s refers to 'a culture called Pilaterode ... and the wood in the same land', so it was an early 'royd' or clearance. 'Pellet' derives from a variety of oats that had husks which did not adhere to the grain and the element occurs in several local field names. One example is Pellet Croft in Kirkburton. See Screamer Wood.

PETER GATE

On the map of 1818 the street that we know as Cross Church Street was called Peter Gate, a name that was clearly inspired by the parish church of St Peter's. At that time it may have been thought of as a possible alternative to Cross Church Street, at least by some people, but there is no evidence that it was ever in popular use.

PHILOSOPHICAL HALL

The Philosophical Society had been formed in 1825, and details of their hall are given in White's Directory of 1837. The Improvement Commissioners had offices in the hall and in 1866 it was sold and converted into a theatre – eventually to be called the Theatre Royal. Huddersfield still had no Town Hall at that time and in 1868 the historian Charles Hobkirk wrote that the Philosophical Hall was one of only two buildings in the town that could be used for public meetings and concerts. It stood in Ramsden Street and was apparently burnt out in 1880, during the running of a play called 'Drink'.

PIG MARKET (Queen Street)

There was little room in the Market Place for the display and sale of animals, so locations in other parts of the town were used for that purpose. The best known of these is the former Beast Market, but the map of c.1780 shows a Horse Fair on the south side of Kirkgate, possibly a temporary site. There was also a Swine Market, dating from 1818 at least, and it catered for the large numbers of pigs that were driven into Huddersfield from as far away as Liverpool. It was located to the south of the Shambles and was also known as the Pig Market. The site later became Victoria Street, until that too was demolished.

PINFOLD

The pinfold was formerly the enclosure where stray beasts would be kept, until reclaimed by their owners on payment of a small fine. Little has been written about Huddersfield's pinfold although the site can be identified and is well documented. In 1627, for example, George Jepson illegally 'rescued' his cow from the pinfold and in 1687 the pinder, a man called Thomas Chadwick, was accused of the false imprisonment of horses that had been grazing on common land. Early 19th century maps show the position of the pinfold, on land close to where Union Street joined Lowerhead Row - today's Old Leeds Road.

PITTS WOOD (Marsh)

At the top end of Greenhead Park there was formerly a field called Cooper Pitts. It is shown on maps from 1716 and is linked historically with the now vanished Pitts Wood. In 1807 this wood was a prominent landmark on the newly-built turnpike road to Rochdale - the New Hey Road - and a toll bar was placed there. The connection was with the Cooper family of Edgerton, and the coal pits were already being worked in the 16th century. A deed of 1580, for example, that had to do with enclosures in the Highfields area, granted Thomas Armitage the right to use a way through the fields 'unto the Coole pittes there aboutes'. The place-name may not have survived the creation of Greenhead Park but it is mentioned in the Waterworks Act of 1876. The Act made provision for the improvement and widening of New Hey Road, near 'the *Junction Inn*' and the 'north-west corner of Pitts Wood'.

PLOUGH INN (Westgate)

The *Plow* is first mentioned by name in 1777, at the Quarter Sessions, and in the Ramsden survey of the following year Sarah Kaye was the landlady. It was located on the south side of Westgate at the junction with Market Street, and is remembered as the meeting place of the founding members of the Huddersfield Choral Society.

PLUMBERS ARMS (Macauley Street)

This is a fine three-storey building in the vernacular tradition, and it may be older than the name. There was no *Plumbers Arms* in the trade directory of 1853, although it is worth noting that a plumber called Joseph Hall was living on Macauley Street. The *Plumbers* may therefore have started out as a beer house. It was named in the directory of 1879, with Elizabeth Ann Iredale as the landlady.

PORTLAND STREET
I have found no explicit reference to the date when Dyke End Lane became Portland Street but from the Commissioners' minutes it seems to have been between 1865 and 1867. The lane ran between Trinity Street and New North Road, an area that was becoming increasingly fashionable at that time, completely altering what had been a rural landscape. No doubt 'Dyke End' had to go because it sounded out of place among such newly-coined names as Belgrave Terrace and Brunswick Place. Portland Street had the right associations for Huddersfield's flourishing middle class in the 1860s. It is surely no coincidence that the family home of the new Mrs Ramsden had previously been the home of the Dukes of Portland, whose family line included a Prime Minister and Margaret, Duchess of Portland, 'one of the richest women in England, friend of George III and Queen Charlotte and patron of botanists and scholars from all over Europe'. See Bulstrode.

POST OFFICE YARD
In c.1800, according to John Hanson, the post office was in Kirkgate, in that part of the street that we now call Oldgate, and it was managed by old Mrs Murgatroyd. The letters were carried round the town by another old woman called Brooksbank. When the post office moved to New Street the location continued to be called Post Office Yard, and both places are shown on the map of 1826. In 1849 the Improvement Commissioners resolved that 'the Old Post Office Yard, leading out of Kirkgate into Castlegate, be paved with 7 inch sets or with old sets and flags'. In 1875 a fine new post office was built on the north side of Northumberland Street and that building survives, directly opposite the present post office.

PRIEST ROYD
The conversion of Priest Royd Mills into apartments named The Melting Pot has restored more than the derelict buildings – it has brought back into popular use one of the oldest place-names in the town centre. The element 'rod', later pronounced 'royd', was used in the 12th and 13th centuries for an assart, a specialist word for land that had been ridded of its trees. Such assarts were often named after the individual responsible, so Priest Royd is likely to have been cleared on the minister's behalf. However, the name is first recorded as 'Preist Roide' in a title deed of 1528, and then at regular intervals through the 16th century. In c.1570, John Bothomley made a further enclosure on 'prestroid banke', and 'Preistroids Lane' was in existence by 1663. In 1853, Thomas William Clough lived at Priest Royd House, a fine villa in its own grounds on the outskirts of the town. He was a solicitor and acted as clerk to the Improvement Commissioners.

PRINCESS STREET (Buxton Road)

This street was probably being planned in the late 1840s for it is on the OS map of 1851 but only partially built. The earliest reference in the Commissioners' minutes is to a stretch of Buxton Road 'between the *Wool Pack* and Princess Street' in 1850, and it is that section that is shown on the map. The Ramsdens had not yet exhausted the stock of 'royal' names but Victoria and Albert were already in use, or about to be so, and the inference is that the 'Princess' was one of the couple's several daughters born in the 1840s. Of these the one most likely to be honoured in this way is the first-born, Victoria the Princess Royal.

PROSPECT HOUSE, PROSPECT ROW, PROSPECT STREET

Prospect House is situated off Outcote Bank, at the east end of Prospect Street, close to the former site of the lodge for Spring Grove. It is a prominent building and has a long history, apparently longer than the street from which it is now named. It is said to have been erected in 1838, as the British and Foreign School, and it became the first headquarters of the Mechanics' Institution. It looks out over the river Colne to Castle Hill, a fine view that almost certainly explains the name. Prospect Row was the name of a row of houses just to the north of the school, on the west side of Outcote Bank, and it was certainly in existence in 1849. Prospect Street was built later, although both names are mentioned in the Commissioners' minutes from 1851. This street lay at right angles to Outcote Bank and was said in 1861 to be separated from the Spring Wood estate by a fence wall. By the 1870s it was a populous street with a square on the west side. This square was named after Robert Watson the builder who occupied no. 5.

QUAY STREET

The surviving section of Quay Street is not linked directly to the town, although Brierleys' mills hint at its former importance. In 1851, for example, it was a busy thoroughfare that provided access from Kirkgate to Turnbridge via the much narrower Rosemary Lane. A Commissioners' minute of 1854 leaves us in no doubt about how busy it was, for it records complaints that the pavements were blocked by 'building materials, timber, iron castings, woollen waste ... loaded wagons and a steam engine boiler'. Its history almost certainly goes back to the completion of the canal, for a way through the fields is shown on the estate maps of 1778-80. However, the name has not been noted earlier than 1816-17, when a directory listed Benjamin Carter of Quay Street as a machine-maker: Richard Clay of 'Quay' was a rope-maker, and there was a ropery at Turnbridge on Crosland's map of 1826. However, on that map the street is called Water Gate, so the place-name may not have stabilised at that time.

QUEENSGATE (Ring Road)

This is the name coined for the southern section of the Ring Road, formerly part of Ramsden Street and Shore Head. It runs between Queen Street and Queen Street South and the choice continued the tradition established by the Ramsdens of using the element 'gate'. It would in turn influence the naming of Kingsgate.

QUEEN'S HEAD

The first Huddersfield inn called the *Queen's Head* occupied a site on the south side of Westgate, with stables and a courtyard to the rear. It lay behind the shops that faced onto the Market Place, to which it had access via a passage. The tenant in 1778 was Thomas Green but he succeeded to a family called Bramhall who had been landlords for generations. The link between this family and the inn may go back to 1680 when a Huddersfield innkeeper called Nicholas Bramhall made his will. The name *Queen's Head* was already traditional at that time, so it probably commemorates Queen Elizabeth. There was a second *Queen's Head* in the town by 1813, located on the north side of King Street, which may imply that the Westgate inn was no longer in business. John Dobson was a banker in the Market Place during that period but he was also a hop merchant and occupied premises in the Queen's Head Yard in King Street. This name has outlived the inn itself.

QUEEN STREET

This is one of the earliest town centre names, recorded in the rentals from 1811. It should probably be linked with King Street, as they were both part of the early 19th century development south of Kirkgate. It is possible that the names were chosen simply because they conferred the right status on the new streets but they might also have been intended as a compliment to George III and Queen Charlotte. The building of the chapel (1819) and Court House (1825) set the tone of the street and that was maintained in the elegant town houses that survive on the east side. The space in front of the chapel was known as Queen's Square. There was a Queen Street on the map of c.1780, part of what would later be Westgate, but that is the only example of the name. It was probably so named because it ran past the Queen's Head.

QUEEN STREET SOUTH

This self-explanatory name is first referred to as an 'intended street' in 1863, in conjunction with Aspley Street, Firth Street and Colne Road. The intention was to improve communications in that part of Huddersfield, in particular to 'provide a direct road from Aspley to the Engine Bridge'. The new route avoided 'a very circuitous and steep road through the town'.

QUEEN'S HOTEL, QUEEN'S TAP YARD (Market Street)

The 'queen' in this case was almost certainly Queen Victoria, not Queen Elizabeth, but the building, with its ashlar façade and cast iron balustrade, dates from before 1830. At that time it was the residence of Mr Joseph Brooke, the Chairman of the Improvement Commissioners, and its gardens extended from Cloth Hall Street to Threadneedle Street. I do not know just when it became a hotel but it is listed in the directory of 1853 with Joseph Gaunt as the manager. The name had changed to the *Queen's Commercial Hotel* by 1879 with Charles Tinker as the landlord. He was also in charge of the 'Tap'. Many Huddersfield people remember the *Queen's* as an inn but it has been a popular restaurant for some years.

Ramsden Street in 1964: the three buildings shown are the Spiritualists' Church, the Public Baths and the Picture House.
Courtesy of the *Huddersfield Examiner*

RAILWAY STREET

The building of the railway station, completed in 1850, opened up the area north of Westgate for development. There was immediately a debate about how best to use the space in front of the station, and it involved both the agents who worked for the Ramsden estate and the Improvement Commissioners. The building of Railway Street and the line it followed was part of the plan that was eventually adopted. The name and the location identify it as part of the New Town of Huddersfield. It is first mentioned in the minutes in 1853 but was still being talked about as a new street seven years later.

RAMSDENS ARMS (Cross Church Street)

This former public house is now 'The Flying Circus', a name unlikely to have been approved of by the Ramsden family. It was an inn used by clothiers in the early 1800s and much public business was also carried on there. In 1835, for example, when the rebuilding of the church tower was in progress, a meeting was held in the *Ramsdens Arms* with the aim of raising money for a new clock, a public amenity that the town then lacked. So many people turned up that they filled 'the large room of the *Ramsden Arms*' and the meeting was adjourned to the yard. Edward Law's history of the inn notes the occurrence of the name in 1795 when Joshua Clegg was the landlord. There are references to Huddersfield innkeepers with that surname as early as 1747, so it may be that this was one of the town's first public houses.

RAMSDEN STREET

This is another of the names that remind us directly of the Ramsden family who held the lordship of Huddersfield for over 300 years. It was almost certainly a name approved by them and it was given to a section of the old Back Green, or Back Green Road, a name still found on the town map of 1818. The Back Green was an ancient right of way for traffic through Huddersfield, one that avoided the town gate and the narrow approach lanes. It would have been particularly busy from 1759 when it became part of the Wakefield to Austerlands turnpike road. There was some building at the top end of the 'street' from the late 1700s but the major development was the erection of Ramsden Street Chapel in 1824-25. The chapel and the street are both named on the map of 1826. John Hanson remembered when 'there was not a single house' on that part of the Back Green, and he claimed to have gathered mushrooms in the field where the chapel was built. This was, of course, where the present library stands.

RASHCLIFFE (Lockwood)

Most of the references to this name date only from the 19th century, when that part of Lockwood was urbanised. As houses, schools and places of worship were built Rashcliffe acquired a status it had never previously enjoyed. The Lockwood historian Brian Clarke found no evidence for the name before 1800 but speculated that it might originally have been 'rush-cliff', a rushy or swampy part of Lockwood common - a view previously expressed by Canon Hulbert. The spelling 'rash', for 'rush', can be found in Scotland and the most northerly parts of England but I have not found it in this part of Yorkshire. Professor

Cooke's warehouse, c.1875: it was purchased by the Corporation for their offices, now the Ramsden Street section of the Town Hall.　Clifford Stephenson collection

Smith suggested a derivation from a dialect word 'rash' that means a strip of uncultivated land and that would be accurate in terms of the topography and land use. Again, it is not a word that I have found in this part of Yorkshire. However, the name is on the map of 1768 as 'Rash Cliff' and I have Rash Cliffe in the charcoal accounts for Colne Bridge 1746.

REAP HIRST (Birkby)
The word 'hirst' is usually associated with wooded and hilly areas and this fits very well with Reap Hirst. 'Reap' is much more of a problem; the word occurs in a dozen or so west Yorkshire place-names, nearly all of them on or close to moorland, a location that rules out some possibilities. The most likely explanation is that it refers to the clearance of scrubby growth, a meaning implicit in one late Nidderdale text. The earliest reference to the locality is in the court roll of 1658 when two Fixby men were accused of digging for stones 'neare the highway upon reape hirste'.

RED DOLES (Leeds Road)
Red Doles is a very evocative name, for it reminds us that this flat part of the valley was where the town originally had its 'ings' - land that was flooded temporarily to produce high-quality grass, good for pasture and meadow. The 'doles' were the shares of individual tenants and they were marked out by stakes or stones. There are one or two ancient names that take the history of the town ings back to the 12th century but, unlike Red Doles, they have not survived. The name is found as 'Reddoles side' in the court roll of 1680 and it may have survived because it later served to identify a lock on the canal. I am uncertain how to interpret the first element in the name but the location might imply that it was the word 'reed'. It seems a little too early to be connected with the 'red' pollution of 'cankered' dykes but, unless much earlier spellings are found, that explanation remains a possibility. See Canker Lane.

RIDDINGS (Sheepridge)
We associate this name now with an area of high density housing and yet in origin it signified land that had been cleared of trees. It was often used as a field name, more commonly in Bradley and Deighton than in other parts of Huddersfield, although most of the evidence is in 18th century sources. In 1756, for example, tenants were forbidden from making a footpath across 'two closes in Furrtown called the Ridings and the Warren': the second of these was almost certainly connected with the name Warrenside.

RIDGEWAYS YARD
I include this name because it is one of the earliest of the Huddersfield yards, listed in 1816. In that year several men were said to be of 'Ridgway-yard' but they were manufacturers, not necessarily residents, and the yard may have been where they carried on business on market days. Among them was a man called John Moore who was probably connected with the Moores of Johnny Moor Hill. The yard may have been named after John Mills Ridgeway, a wool stapler.

RIDING SCHOOL (Queensgate)
I am uncertain just when the Riding School was built, although the date 1847-48 is given in several sources. It was used initially by the West Yorkshire Yeomanry Cavalry and originally had a tall arched entrance with bas reliefs of rampant horses to each side. However, the building was later given several extra storeys, possibly about 1905, and it became a variety theatre, aptly called the Hippodrome. In December 1967, when it was the Essoldo cinema, it was partially burnt down: the renovated building was reduced to its original height and renamed the Classic. More recently it has become a bar and, with the cladding removed, we can again see the horses that remind us of its earliest history. The distinctive Italian Palazzo style that characterised it when it was new can also be seen in the adjoining *Zetland Hotel.*

RIFLE FIELDS
This appears to be a relatively recent name and yet I am unsure of its precise origin. Even in 1901 there were open fields in this part of the town and that did not change until Park Avenue was extended to the south. The likelihood is that it was where the Huddersfield Rifle Club or Corps had arms practice, although their drill hall between 1863 and 1901 was in the Riding School in Ramsden Street.

RING ROAD
In the 1960s it was judged that a radical redevelopment of the town was necessary, to take account of the increasing traffic, improve shopping facilities and give the central area 'real life'. The Inner Ring Road was seen as a key part of this plan, at a time when the M1 and M62 were still not completed. The optimism felt by many people in those years was expressed by Roy Brook in The Story of Huddersfield (1968). The enormous changes that the new road eventually brought to the town emerge in this present book in the catalogue of demolished streets and buildings. A retrospective view in The Buildings of Huddersfield (2005) makes the following points:

> *... the ring road ... when completed in the 1970s, achieved the aim of excluding traffic from the town centre but at a very high cost. It isolated the centre from easy pedestrian access to the surrounding areas and obliterated the clear relationship between the town centre and nearby residential streets. The superb vista up New North Road was destroyed and St Paul's Church was arbitrarily cut off from the end of Queen Street'.*

ROSE AND CROWN
The old *Rose and Crown* was apparently demolished in 1884, but photographs of it have survived. It stood on the south side of Kirkgate, in the same block of property as the vicarage, and was originally church property. It is uncertain just how long it may have been an inn but Edward Law found a reference to it by name in 1773, in connection with local carriers. Thereafter it is well documented, although it was in decline by the late 1840s. The name is a reference to the royal coat of arms.

ROSEMARY LANE

What survives of this street, at the bottom end of Kirkgate, is only one side of the original lane. It was formerly one of several very narrow alleys that developed on the Low Green, probably in the 1700s. It is difficult to say how old the name might be, even though the line of the lane can be seen on the estate maps of 1778-97. It is certainly found in trade directories from 1809 and on maps from 1818, but it is the map of c.1780 that may hold a clue to its origin. On that map there are several names that seem likely to have been in common use but had no 'official' status, e.g. Midwood Row and the Maze. The lane we know as Rosemary Lane is shown there as Sh(i)tten Lane, perhaps because of the Beast Market near by. I imagine that Sir John Ramsden would have preferred to associate the street with 'the strong, sweet smell' of the shrub Rosemary!

ROW

This was the popular word for a line of houses, apparently more common in the North and Scotland than elsewhere. The 'Raw' in Huddersfield, first recorded in 1716, was a line of houses at the bottom end of Kirkgate, on the north side, and the name is found in rentals well into the 19th century. It may have influenced the choice of Lowerhead Row and Upperhead Row, although the element occurs in a number of other names from that period, e.g. Cropper Row and Midwood Row.

SADDLE INN

The *Saddle* was formerly on the south side of Westgate, with the brewhouse to the rear, and the name is on record from c.1800. It is one of the few inns where we can be reasonably certain that the name had its origin in the landlord's main occupation. Richard Thewlis held the licence in the early 1800s and on several occasions he was described as a saddler. Thomas Armitage was the landlord in 1837 but the name did not survive for very long after that.

SAINT GEORGE'S SQUARE

The open space in front of the railway station was being called St George's Square soon after the completion of the new *George Hotel*, a name which incidentally supports the origin offered for the older *George Inn*. The railway line between Huddersfield and Manchester had been opened in July 1849 and just a few weeks later Alexander Hathorne wrote to his superior George Loch, suggesting that the land 'in front of the New *George* ... should be thrown open and formed into a Square', in order to show to advantage the 'beautiful Railway Station and still more handsome ... New *George Hotel*'. The early history of the Square, and the involvement of both the Ramsden family and the Improvement Commissioners, was well told by David Clarkson in his article in the local history journal Old West Riding (1989). White's Directory of 1853 comments in glowing terms on the hotel and other 'elegant stone buildings' in 'St George's Square', all recently erected.

SAINT JOHN'S CHURCH, SAINT JOHN'S ROAD

St John's Church was paid for by Sir John William Ramsden and consecrated in 1853. It was part of the New Town that followed on the development of the railway station and the new *George Hotel*, an out-of-town location that was

Alfred Jubb's Albany Hall and Printing Works, St John's Road, 1905; named in honour of the Duke of Albany. The Hall was the former Collegiate School and the works, built in 1891, reflect aspects of its architecture. Clifford Stephenson collection

linked to the building of John William Street and its northward extension. This road replaced a much earlier footpath to Bay Hall, Birkby and Storth that passed through the Swan Yard. It is first referred to in 1856, as 'a proposed new road ... from Fitzwilliam Street to the limits of the Improvement Act, near Bay Hall Church' but after the church dedication it soon came to be known as St John's Road.

SAINT PAUL'S CHURCH, SAINT PAUL'S STREET
St Paul's Church was one of the so-called Waterloo churches, and it was built in 1829-31 on land donated by Sir John Ramsden. It was located on the Back Green, across from the end of Queen Street, and later became the focus for residential housing. In 1838, the Lighting Committee placed a lamp 'on the Back Green' half way between Commercial Street and the church and this was followed by a lamp in St Paul's Street in 1842. The street ran southwards from the east end of the church, passing between the Drill Hall and the Technical College. It survives in name as part of the University campus.

SAINT PAUL'S GARDEN
This was a small, triangular oasis of green in 1907, located roughly where the main entrance to the University now is. On a photograph of that period it can be seen as an open space, planted with trees and hedges; children are playing games there whilst their elders relax on a bench. The 'Garden' may have had its origins in a low-key dispute between the Commissioners and Captain Graham, the local agent for the Ramsden estate. In 1867 the agent claimed that 'the line of St Paul's Street had been turned so as to correspond with the end of Zetland Street', in an unsatisfactory manner and without consultation. He later suggested that the Commissioners should pay for re-aligning and paving the street, leaving a small, triangular piece of ground that would 'be fenced off ... and planted by Sir John'.

SAINT PETER'S STREET
It takes its name from the parish church and is first mentioned in the Lighting Committee minutes in 1847, although it can only have been an 'intended' street at that time. On the map of 1851 it was very short, and stretched only from what is now Lord Street to Northgate. The extension westwards, eventually reaching beyond John William Street, was part of the New Town development after 1850 and ten years would pass before the work was completed. It seems from the map of 1850 that there may have been a footpath at the north end of the churchyard known as St Peter's Lane.

SAUGHALL (Bradley - lost)
The few references to this locality occur in the 1630-60 period, and the inference is that it was somewhere near, if not the nickname for, Bradley Hall. The first element may be from a word for the willow tree, found also in Saughes Farm (Mytholmroyd), or even from the dialect word for a gutter or drain. For example, that word is found in the locality named Sough Hall in Rotherham. There is one other consideration. In Bradley it was not unusual for 'hall' and 'hole' to interchange and it may be worth noting that Michael Woodhead was ordered in 1665 to 'open his soughholes to vente the water'. This was part of the irrigation system in the town's open fields.

SCREAMER WOOD (Bradley)

I can offer no explanation of this name which I first note on the OS map of 1843. It apparently replaced the older name Pellet Royd Wood soon after 1829, at a time when some other Bradley woods were renamed. Perhaps the new owners chose to do that as a mark of their independence. See Pellet Royd Wood

SEED HILL (Shore Head)

This is a genuinely old name. It is mentioned in the court roll of 1662 when Edmond Shaw failed to repair 'the common way leading from the place called Seedhill to the place called Churchyate'. This was almost certainly a reference to part of Kirkgate, from Shore Head as far as the 'yate' or gate into the churchyard. Huddersfield's first grammar school was built at Seed Hill soon afterwards, although the endowment deed called it Miln Hill on that occasion. We know that Miln Hill took its name from the corn mill, so Seed Hill may point to a connection with the tithe barn that stood near by. The Commissioners' minutes show that in the 1840s the town's horses were stabled at Seed Hill.

SERGEANTSON STREET

In 1849 the Commissioners drew the attention of the Trustees' agents to the state of 'the intended … street called Sergeantson Street', saying that without sewers it was impossible to drain the new 'Water House' there. The street was named after Mr Sergeantson who was appointed as a trustee of the Ramsden estate in 1844. The development of the street was contemporary with that of Dundas Street.

SHAMBLES LANE (Old Market Hall)

Shambles Lane is an evocative name, since it recalls a part of Huddersfield that the planners and developers did away with – much to the dissatisfaction of most of the inhabitants. It was a picturesque part of the old town and nostalgic memories of it still find their way into the columns of the local paper. And yet those memories have more to do with the Market Hall and the shops on Shambles Lane than with the old Shambles. This building was removed to make way for the Market Hall in 1878-80 but it had previously been at the top end of King Street, close to the junction with New Street. As these were referred to as 'the New Shambles' in 1771, there must have been an even earlier site. Their history is likely to go back to 1671, when the town was granted market rights, but I first find the name in the court roll of 1723. Etymologically 'shambles' is the diminutive of a word for a bench. It came to mean a place where goods were exposed for sale and then particularly a butcher's stall and more generally the meat market. In 1853 no fewer than thirty-four butchers had shops in the Huddersfield Shambles.

SHEAR'S COURT

There was a public house called the *Shears* in the Beast Market, and Edward Law noted that Enoch Booth was the licensee in 1827. Pictures of it as a half-timbered house survive from the early 1900s, shortly before its demolition. The reference is almost certainly to the shears used by cloth-makers.

SHEARING CROSS (Fartown)

Philip Ahier linked Shearing Cross with other 'cross' names in the Huddersfield area, such as those at Cowcliffe, Marsh and Sheepridge, but I have found no reference to it earlier than 1822, in Baines's directory, and no symbol of a cross in that location on any of the Huddersfield maps. The map of 1780 has a field called Mag Croft at this point, cut through by what was then the new road to Halifax. On the OS map of 1851, Shearing Cross was apparently the name of a small triangle of land north of Willow Lane. The triangle, in which there were by then several buildings, was formed by Hillhouse Lane, the Old Halifax Road and the new Bradford Road. The name has been preserved in a new development near by.

SHEEPRIDGE, SHEEPRIDGE CROSS

Sheepridge is a self explanatory name, for this was a hilly and extensive ridge of common land where sheep were grazed. The name is likely to have been coined early in the township's history but the first reference that I have noted is for 1521 when clearance was taking place on 'the common waste called Shepperygge'. This process may have started with Thomas Brooke's 'new house', but it continued over the centuries as Huddersfield's population increased. Other families said to be 'of Sheepridge' in the 1600s were the Barkers, Hudsons and Mellors, and the survey of 1716 provides evidence of many more. The cottages of these tenants were scattered, some of them in locations named as Brakenhall, Browside and Sheepridge End. The highway from Huddersfield to Leeds passed over the common at that time, and Sheepridge Cross was a prominent landmark. It featured in a case at the Quarter Sessions in 1739, when the inhabitants of Deighton were ordered to repair the highway, and several times after that as the dispute rumbled on. It is also mentioned in the accounts of the surveyors of the highways. In 1789, for example, 'causeway stones (were) laid towards Shepridg cross'. The cross marked an important junction on the common, roughly where Wiggan Lane now joins Sheepridge Road. This highway ran northwards, through Bradley Gate, Shepherd Thorne and Toothill.

SHEPHERD THORNE (Bradley)

The 15th century reference to this name listed by Smith is incorrect. The document he quoted had to do with the pasturing of sheep in a locality called Shepherdthorn, but one on the commons between Notton and Darton, near Barnsley. The first reference I have found to the Bradley farm is in a deed of 1557, when Thomas Brouke was said to be 'of Shepperdthorne in Nether Bradleye'. Other surnames linked with the farm in the next one hundred years were Hirst, Savile, Rawnsley, Fox and Scott.

SHIP INN

For over 200 years the *Ship Inn* stood near Shore Head, on the north side of what became Queensgate. It served the local community as a public house and was a place where societies could meet and inquests were held. Among the early landlords were Joseph Mappleson, James Carleton and Frederick Richardson. The name may have had a specific origin, for Edward Law notes that a Huddersfield innkeeper called John Wilson was the owner in 1794 of 'a vessel or keel with the mast and sails'.

SHORE, SHORE BANK, SHORE FOOT, SHORE HEAD

The important element in all these names is 'shore' which is usually explained as a steep slope. That may not be an accurate enough definition, for David Shore's research suggests that the 'shore' was an arc of rising land above a river or stream. The name is not uncommon in places near to highways that cross a water course. The Huddersfield locality is first mentioned in connection with the Huddersfield corn mill, in an undated early 16th century deed. This is referred to later as Shower Mill (1584) and Shoarefoote mylne (1611). The name Shorehead may not appear on any street sign but it is still in everyday use in the town, used now of the area close to the roundabout above Somerset Bridge. This name actually goes back to 1563, when 'Shoreheade' was a dwelling, in the tenancy of a man called John Horsfall. In the 1860s Shore Bank was the name of a short street that branched eastwards from Commercial Street, now part of the University campus.

SHOULDER OF MUTTON

Proverbially these words signified something that was good and wholesome, and they also have a long history as a 'pub' name. Several reasons for this popularity have been offered, and it may have been a traditional choice when the licensee was also a butcher. In 1777 a public house called the *Shoulder of Mutton* stood in a detached group of buildings on the north side of Temple Street and Richard Blackburn was the landlord.

SILK STREET

I am puzzled by this name. The street is shown on the town map of 1818 as a narrow way to the east of the Beast Market, linking Lowerhead Row and Rosemary Lane. Its history through the 1800s is well documented and it survived as a short thoroughfare into the 20th century. The puzzle is to determine what lies behind the name, especially as it appears to date from the time when name-giving was largely controlled by the Ramsdens. I can find no obvious precedent in other towns and cities and no local family or individual called Silk. One or two silk merchants had premises close to Huddersfield so perhaps there had been a retail outlet in the street.

SILVER STREET

This is a popular English street name, recorded from at least the 14th century. It is shown on the Huddersfield map of c.1780, as a section of the town street, roughly where Venn Street now is. There is no other evidence for the name and the suspicion is that it was chosen by the map maker, like Chancery Lane, for its prestige value. The Silver Street in Aspley is from a later period.

SMITHSONS

By the 1600s the oldest Huddersfield settlements all had distinctive names, many of them topographic in origin. Typical examples are Croft Head, Oak and Clough House. On the other hand, it was common for more recently built houses to be known by the name of the family in occupation, and examples already mentioned are Boothroyds and Clays. 'Smithsons' is first mentioned in 1667 when a family called Shaw was living there but it seems certain to have been 'the house of Edmond Smithson' mentioned in the court rolls in 1649. It was somewhere close to the Town Head but the name did not survive.

SMITHY HILL

I have references to a dwelling house of this name between 1589 and 1657, occupied at different times by families called Shaw, Armitage and Brooke. Although its exact location is unknown, the documentary evidence suggests that it was somewhere on the town gate, the site of a former smithy. It may have disappeared when the Market Place was developed after 1671.

SNODLEY

This place-name is seldom used now but the hill in question was formerly the site of a service reservoir, at the junction of Trinity Street and Mountjoy Road. There are now houses on that site. The name has a similar run of spellings to Snowden Hill in Hunshelf, near Penistone, and could therefore be explained as 'bald hill', derived from the Old Norse word 'snothin'. In 1436, a grant by Richard Beaumont to William Couper of Edgerton referred to a dyke 'by the east side of Snodenhyll' and a lease of 1512 mentions 'closes called Snawdunhile'. Later spellings include Snoddle Hill (1716) and Snodley (1854).

SNOW HILL

In the opening years of the 19th century, the name Westgate applied only to the stretch of road from the Market Place to Market Street. At that point it parted, to pass through the 'Maze' of buildings that was known as the top of the town. The branch to the south, roughly parallel to Half Moon Street, was called Snow Hill. The name is shown on maps of 1818 and 1826 but did not survive the development of the area soon afterwards. There may have been some influence from Snodley, further to the west, but it is more likely that London's Snow Hill served as a precedent. The same name survives in Birmingham.

SOMERSET BRIDGE

A debate about the need to rebuild Huddersfield Bridge began in the 1860s and it was finally demolished in 1872. A temporary wooden structure was erected alongside it and the new bridge was completed by 1874. Lady Guendolen Ramsden performed the colourful opening ceremony and, since she was the youngest daughter of the Duke of Somerset, the bridge was given that name in her honour. Close by are the Somerset Arms, Somerset Crescent and Somerset Road.

SOUTH PARADE

The word 'parade' originally meant a show or display but by the 17th century it was being used for a display of troops and also for the places where they mustered or marched. As a name it could be applied to public squares, promenades and streets, especially those with rows of shops. Among the best known names of this type are Grand Parade in Bath and Kings Parade in Cambridge. South Parade was the earliest Huddersfield street to be so called, recorded in the directory of 1822 and shown on the map of 1826. The Commissioners had their offices there from 1848, ending a long tradition of meetings in the *George Inn*. It was another of the streets that were lost to the Ring Road.

SOUTH STREET
The earliest part of South Street was built c.1840 and it ran south from Trinity Street (or West Parade as it was then) to Spring Street. A petition asked for street lamps in January 1842 but the matter was deferred by the Lighting Committee. Ten years were to pass before plans were put forward for an extension of the street to the south, and when this was completed it stretched all the way to Outcote Bank. It was part of a grid plan of new streets, with George Street, Henry Street, Swallow Street and Prospect Street to the east. In 1861 it reached 'the boundary fence of the Spring Wood estate'. The Ring Road and car parks now occupy most of the area where those streets were, although the name Old South Street survives. It is about the same length now as it was in 1840.

SPARROW PARK (George Street)
See Vagrant Office.

Sparrow Park, 1969; at the junction of Springwood St and George St, directly over the railway tunnel. It gave way to the Ring Road. Courtesy of the *Huddersfield Examiner*

SPREAD EAGLE

This is another traditional 'pub' name, recorded from the 16th century at least. It derives from the heraldic charge which shows the eagle looking to its right, with its wings spread out and legs extended. The Huddersfield *Spread Eagle* was on Manchester Street, and it is first referred to in 1778 when William Wilson was the tenant. At that time it was surrounded for the most part by crofts and gardens, and there was a large bowling green to the north. Near by were a few cottages tenanted by Joseph Hawkyard, possibly the ancestor of Abraham Hawkyard, the landlord of the *Spread Eagle* in 1853.

SPRINGDALE AVENUE, SPRINGDALE STREET (Longroyd Bridge)

Springdale is now thought of as a small part of Thornton Lodge but the name first seems to have been given to a villa on the south side of the Colne, close to Longroyd Bridge. It was the home of a merchant called Wrigley in the 1850s and a family of solicitors called Fisher in the 1870s. When the villa was demolished the area was given over to housing and the present streets commemorate its name. The Starkeys' mill at Longroyd Bridge, now also demolished, came to be known as Springdale Mill.

SPRING GROVE, SPRING GROVE STREET

This was an early and prestigious mansion, shown on the Lister Kaye map of 1791 as Mr Fenton's 'New House'. It may already have been given its name, for Spring Grove is listed in a directory thought to have been compiled in 1790. A letter dated 1792 describes William Fenton as 'of Spring Grove'. The Whig politician Lewis Fenton was in residence there in 1832 and his opponents are

The Grove public house, Spring Grove Street. William Fenton named his villa Spring Grove, c.1790, and it was the source of both names. G Redmonds

*The precipitous steps that lead from Bankfield Road to Spring
Grove; one of the routes up the steep bank.*   G Redmonds

said to have broken almost every window in the house. In 1857, the trustees
of his estate described it as a property that was difficult to let and expensive
to maintain. It was in the occupation of the solicitor Edward Lake Hesp at that
time but not, they said, 'at a fair rent'. They felt themselves compelled to take
what they could get, and estimated its value as no greater than 'an ordinary
tradesman's house'. It was finally demolished in 1879 to make way for Spring
Grove School. The fact that 'grove' meant a small wood suggests that the name
was inspired by Spring Wood, just a few hundred yards away. It was a significant
choice. Although 'grove' would have been a familiar word to Mr Fenton, it was
not used in the Huddersfield area where 'greave' had much the same meaning.
Nor is there any previous history of 'grove' as a villa name locally, so Mr Fenton
was something of a trend setter. Subsequently, numerous villas in and around
the town were to have 'grove' as their second element, and these influenced
other minor names, e.g. Edgerton Grove and Edgerton Grove Road.

SPRING LODGE (Longroyd Bridge)

This was a fine villa in the early 1800s, located on the steep slope above Longroyd Bridge, and the name is further evidence of how the elements 'spring' and 'lodge' influenced local place-names. The merchant John Starkey was living here in 1837 and Lewis Starkey M.P. in 1879. The house still stands, shorn of its former grandeur but a reminder of the wealthy and influential Starkey family.

SPRING STREET

The Waterworks Offices were built in 1828, and the first houses on Spring Street date from that period, running west from Upperhead Row. They were elegant terrace houses, built of fine ashlar stone, and were clearly intended for the town's growing middle class. Recent refurbishment has helped to restore something of that status. Sir Charles Sikes, the man who started the Post Office Savings Bank, lived here, and on the south side of the street is the former infant school where the Choral Society had its first quarterly meeting in 1836. The original and eastern ends of Spring Street and Back Spring Street, where the Bus Station now is, were demolished in the 1970s to make way for the Ring Road.

Back Spring Street c.1970. It was demolished when the Ring Road was built.
Clifford Stephenson collection

SPRING WOOD, SPRING WOOD HOUSE

There are several early place-names in Huddersfield with 'spring' as an element. These include Springnok (1533), Springe (1591) and Spring Head (1661). However, without additional information we cannot say whether these referred to water sources or to coppice woods. Spring Wood in Marsh, remnants of which survive, seems likely to have been a coppice, for this was the word used locally to describe such woods. However, the map shows a 'spring' in the corner of the wood so we cannot be absolutely certain. In 1804 a manufacturer called Joseph Haigh bought land here from the Lister Kaye family and built himself a prestigious mansion. It came to be known as Spring Wood Hall but was demolished to make way for a housing estate in the 1960s. Directly and indirectly it was to be responsible for numerous Huddersfield place-names. Among these are several street names on the estate or close to it; Spring Street, Spring Wood Avenue and Spring Wood Terrace, Spring Place off Upperhead Row and all the names associated with Spring Grove.

SPRING WOOD STREET

Trains leave Huddersfield for Manchester by a tunnel that takes them underneath the top end of the town and emerges at Spring Wood in Marsh. Spring Wood Street and the ventilation shafts mark exactly the line that it follows. In fact, when the street was first planned, in 1849, it was called Tunnel Street and this name survived for almost twenty years. However, in 1868, Sir John William Ramsden agreed to change the name to Spring Wood Street in response to a complaint by the residents. See also Bow Street.

Springwood Street; the ventilation shaft for the railway tunnel. In the background is the factory of Conacher and Co. - specially designed for building organs. G Redmonds

STABLES STREET (Chapel Hill)

The first reference to this street, in the Commissioners' minutes, solves the meaning of the name. In November 1849, an application was made to erect a building on a plot 'adjoining the intended new street to be called Stables Street, leading out of the Chapel Hill below Messrs Stables' warehouse'.

STANTON'S THEATRE

John Hanson said that in the early 1800s there was a theatre in the Old Post Office Yard, converted out of an old 'laith' or barn. John Stanton 'the celebrated scene painter was the proprietor' and the building 'held a large number of people'. Little has been written about either Mr Stanton or his theatre but the rental of 1798 includes a reference to both, and the site is marked by the word Playhouse on the map of 1818. In 1820 this had been shortened to 'Play'.

STAR INN

This popular pub name is said by writers to go back to the 1400s, and the allusion is likely to be either heraldic or religious. Edward Law notes that a man called John Coats occupied a cottage at 'Huddersfield Bridge end as a public house' in 1796, and this was the location of the *Star* in the early 1800s. George Robinson was the landlord at that time and he was followed by James Brook and then Joseph Brook. The latter worked for the Ramsdens and was a prominent surveyor in the town.

STATION STREET

The completion of the railway station, in 1850, prepared the way for further development in that part of Huddersfield, and buildings at the Westgate end of Station Street are shown on the town map of 1850. Progress may have been interrupted during the debate over St George's Square but nonetheless the street features regularly in the Commissioners' minutes from the 1850s. It was, of course, part of the New Town, and the last great project was Station Buildings which occupies the corner site at the junction with St Peter's Street. Work here is said to have been completed in the early years of the 20th century.

STONE DELVES (Netheroyd Hill)

Small quarries were called 'delves' locally and this place-name is further evidence of early stone-getting in the Netheroyd Hill area. It is first mentioned in a title deed of 1553, in connection with Longwood House, and there appear to have been two dwellings there. The parish registers carry details of Edward Brooke 'of Stonedelfe' from 1563 and Robert Batley 'of Stone Delves' from 1578. A survey of the Ramsdens' property, in 1625, lists 'a messuage ... called Stonedelves' and another 'messuage called allsoe Stonedelves'. The tenants on that occasion were Grace Brooke and George Batley. The houses were located on the north side of Netheroyd Hill Road, just below the *Shepherd's Arms*, but I have no evidence that the name is still in use.

STONY LEE (Birkby)

The burial of Edward Brooke 'of Stonyelee' took place in 1566, and shortly afterwards his widow was granted an acre of land 'on the sun side of Stonylie'. It was a relatively small area of common but is clearly shown on the map of 1716, just to the north of Clayton Dike. In 1718 a dispute at the Quarter Sessions had to do with a footpath that led from the town 'to the common pasture called Stony Lee'. The name seems to have fallen out of use but the locality is remembered in the names Clayton Fields and Lee Head.

STORTH, STORTH LANE, STORTH ROAD (Birkby)

This place-name derives from a Scandinavian word and occurs several times in the Huddersfield area. It is usually said to mean 'a young wood or a plantation growing with brushwood' but the Huddersfield evidence points more to scrubby tree growth than to plantation. This meaning would fit the assart or clearing called 'Storthe' mentioned in a Fixby deed of 1286. Indeed, this reference may be the first to Storth in Birkby, since the township boundary in that valley was in dispute over a long period. The farm called Storth, which was demolished some years ago, is well documented from 1562 when Roger Shaye was the tenant. It was later partitioned, possibly in the early 1600s when the Brookes and Hirsts were living there. In the 1700s the Haighs and Gambles were the tenants. However, the place-name described a 'territory' rather than just a dwelling house: in c.1570, for example, Thomas Brooke enclosed an acre of land 'upon one place commonly called the Storthe'. Storth Lane survives as a continuation of South Cross Road and the name is also preserved in Storths Road and Storth Place.

STORTH (Dalton)

This was the name given to an extensive tract of land between Somerset Bridge and Mold Green, mostly to the east of today's Wakefield Road. Much of it was probably covered in scrub in the Middle Ages but it had become an enclosed pasture by 1483, tenanted by Thomas Blackburne. This family sold the land to the Ramsdens in the late 1500s and it was then incorporated into their lordship of Huddersfield. Eventually, as both Aspley and Mold Green became built-up areas, Storth too gave way to housing and industrial development. The Tolson family had a mill there from the early 1800s, and their name is remembered in Tolsons Yard across the road. Shaw's Pickle Factory now occupies the mill site and some of the present buildings may survive from that period.

SWALLOW STREET

This name is not straightforward. It may commemorate somebody called Swallow, for the surname was well known in the neighbourhood, but I have been unable to identify a likely individual. It is mentioned as early as 1822 but it has the look of an 'intended' street on Crosland's map of 1826 and work was still in progress there as late as 1860. In fact, it was part of the development in the area that adjoined the Spring Wood estate and it followed on the building of Upperhead Row. Now only the name survives, on a forlorn sign in the car park close to Castlegate. Swallow House near by is a rather more substantial reminder of its existence.

Summer in the Grimescar valley, c.1909; an evocative picture of Storth, with Oaklands on the skyline, one of the new villas in Birkby.   G Redmonds collection

SWAN or WHITE SWAN

This is one of the oldest recorded pub names in the country, known since the 14th century when the swan was a popular heraldic device. It was used, for example, by Edward III and later by Henry VIII. Edward Law has shown that the *Swan* in Huddersfield was one of the oldest public houses in the town, mentioned in John Turner's diary as early as 1732. The landlord until 1754 was Thomas Shepley but soon after his death his daughter Elizabeth married Thomas Dransfield and their descendants were still at the *Swan* in the early 1800s. It is also recorded as the *White Swan* from the 1780s. See Swan Yard.

SWAN WITH TWO NECKS

This pub name is on record from the 16th century but the Huddersfield inn was first noted in 1803, at the bottom end of Westgate on the south side. Since there is no obvious reason for speaking about a two-necked swan the explanation offered is that 'necks' was a reference to the 'nicks' that were made on the swan's beak, in order to demonstrate ownership. The licence was held by members of the Clayton family for over fifty years but, in 1862, it was proposed to demolish the inn and 'offer the site so cleared for the erection of a new bank for the Halifax and Huddersfield Banking Co.'. That is what happened but the same name was given to another inn close by, and this is the site of the present *Royal Swan*.

SWAN YARD

The stabling for the *Swan Inn* was formerly on the north side of Kirkgate, roughly where Byram Street now is. It is first recorded as Dransfield's Yard in 1798 when Thomas Dransfield was the landlord of the *Swan*. By the 1830s this yard was the location for a number of small businesses and it must have been an active and atmospheric place. There were blacksmiths, a joiner, a cabinet maker, a coach builder, a chimney sweep and a painter; Wood and Walker were woolstaplers; Nutter and Barrot cotton warp manufacturers. The only link with the yard's former function were the livery stables, operated by John Sunderland. This was not a street but a long, narrow yard, traversed by a footpath to Bay Hall and beyond. Although it was quite out of keeping with the New Town development of c.1850 it would be the late 1870s before the final demolition took place and the building of Byram Street could begin.

SWINE MARKET See Pig Market.

SYKE HOUSE (Lost)

The rectory of Huddersfield belonged to Nostell Priory and it was sold after the Dissolution to two speculators called Andrews and Chamberlain. The deed dates from 1543 and among the items listed in the inventory of the property were 'a cottage ... in the tenure of Thomas Sykes of Huddersfield' and 'a cottage called Sykehouse ... in the tenure of Richard Horsfall'. This surname and the suffix 'house' may point to a link with Fartown but nothing more is known about the name which is mentioned again in 1595. It is noted here simply to complete the list of early place-names.

TEMPLE CLOSE, TEMPLE STREET (Westgate)

Temple Street is first referred to in the directory of 1814-15 when a family called Midwood lived there. It was probably a later name for Midwood Row, shown on the map of c.1780 as a narrow street or alley at the top end of the town. According to John Hanson, the popular explanation of the name was that a house opposite the *Plough* was locally called Solomon's Temple, in remembrance of Dr Solomon, a celebrated quack doctor who lived there. He apparently sold a 'nostrum' which he called Solomon's balm of Gilead. There is some support for this story in the minutes of the Lighting Committee: in 1827 they resolved 'that the lamp upon Solomon's Temple be taken down'. The name is commemorated in Temple Close, just off St George's Street.

THOMAS STREET

This was a comparatively short street between Northgate and Lowerhead Row, roughly where Lonsbrough Flats now stand. There were enclosures on the site in 1826 but the street name is recorded in Whites' directory of 1837 and two properties listed were John Mather's academy and George Bradley's beer house. It may have started out primarily as a residential street but the position had changed by 1879 when most of the buildings were occupied by tradesmen. The residents at that time included beer retailers and shopkeepers, a grocer, a greengrocer, a baker and a bill poster: the only professional man was William Dufton, a dentist. Whilst there were eight Commissioners called Thomas, the street is possibly named after Lady Isabella Ramsden's father. See Dundas Street.

THORNTON LODGE (Lockwood)

The Thornton family moved into the Colne Valley some time before 1545, probably from the Bradford area. They had the fulling mill at Paddock Foot in Elizabeth's reign and were prominent clothiers in that part of Huddersfield for centuries. Edward Law's research into Thornton Lodge has demonstrated a clear connection with this family, particularly with Richard Thornton who was a merchant in Hamburg. This man had inherited an estate near Longroyd Bridge in 1766 and he bequeathed it in 1790 to his son John, also of Hamburg. John sold the property at the *George Hotel* in 1791 and it came eventually into the possession of John Horsfall, a member of the family whose house in the Beast Market was called the Well. This was a period when many fine mansions were being built, and notable examples locally were Spring Grove and Heaton Lodge. John Horsfall clearly wanted something to rival these two fine houses and for the task he chose the local builder Joseph Kaye. He called it a 'lodge' because this was becoming a fashionable name for a mansion but chose also to commemorate the family from whom he had purchased the land. The house has survived, but we need to look at plans and maps to have any idea of its previous status. An early writer described it as 'beautifully situated on gently rising ground and nestled among trees' – an idyllic rural setting – but, as the valley was industrialised, the estate declined in value and was finally sold in 1886. The development that followed created the district of Thornton Lodge that most people are now familiar with – an area of high-density housing.

THORPE HALL (Marsh)

In 1611 Richard Beaumont of Whitley leased the moiety or half of a messuage called Thorpe Hall to Roger Heaton, a clothier. The only other evidence that I have found for the house dates from the same period, and its exact location is not certain. We can deduce that it was probably in Marsh, since Roger Heaton was elsewhere described as 'of Marsh', and it may therefore have been close to Heaton Fold. We still cannot explain the place-name although both the elements have obvious etymologies. Thorpe, for example, is known locally as both a place-name and a surname: the suffix 'hall' may point to an important building but the evidence we have is confined to the few place-names mentioned.

THREADNEEDLE STREET

This name is listed in White's directory of 1837 and mentioned in the Lighting Committee minutes from 1845. However, it has never been an important street, either commercially or residentially. In fact it is no more than a narrow alley, linking Market Street to Upperhead Row, squeezed in between Macauley Street and Dundas Street. It may seem to have been aptly named but the inspiration must be London's Threadneedle Street, the location of the Royal Exchange. Historians are still unsure whether the name is a reference to the three needles of the needle-makers, or to a children's game called 'thread the needle'.

TOP GREEN

During the 1700s much of the 'waste' at the top end of the town, originally called the Over Green, was enclosed, and many of the fields bore the tell-tale name 'Intake'. It is rather surprising therefore to come across the name 'Top o'th' Green' as late as 1822. The inference is that by then it was the name of a house, or a small group of houses, and not a name for what survived of the common.

TOP OF BANK

See Bank Top.

TOP OF THE TOWN

See Town Head.

TOWN

From a very early date 'town' was the word used locally for what we now call a 'village'. It may have indicated that there was a nucleus of houses but it did not in itself imply an urban settlement. A family called 'del Tone' or 'de Toun' was living in Huddersfield from 1285 and this may be evidence of a cluster of dwellings, possibly near the church. Over the centuries the word served to define numerous localities within Huddersfield and some of these, such as town gate and Town Head are dealt with individually. Other examples, with the earliest dates noted, were Townebrowke (1573), Town ditch (1740), Towne end (1689), Towne Lone (1537), Towne streete (1657), Town well (1776) and Towne ynges (1545). Some of these may have been genuine place-names but on the whole they had the value that 'village' has in a phrase like 'the village street'. See Fartown.

TOWN GATE

It may seem confusing that 'gate' should have two apparently contradictory meanings. As a word of Old English origin it signified door or opening, and in the past this was usually spelt 'yate' in west Yorkshire. For example the gate into Huddersfield churchyard was called the 'churchyate' in 1662. As a Scandinavian word 'gate' signified street or road, and Huddersfield's main street was originally known as the town gate. It traversed the town on an incline, rising gently from east to west. In 1589, for example, Dorothy Blackburn's house was said to butt 'upon the town gate in Huddersfield'. Strictly speaking this was not a place-name, simply a description of the street's status, and it gave way to the more fashionable 'town street' in the 17th century. Later still it would be given the name Kirkgate by the Ramsdens, as they sought to enhance Huddersfield's status and have it thought of in the same way as its neighbours Wakefield and Leeds.

TOWN HEAD

The term Town Head was in regular use from the mid-17th century, and it occurred occasionally as late as 1811. From the early 1700s it seems to have been more or less synonymous with phrases such as 'the top of the town' and these two expressions overlap for a considerable period. It is probable that they did not have the same meaning initially. In 1639 Jane Midwood was said to be 'of Towneheade', and the inference is that this was a single house, one that was literally at the head of the town gate. By the early 1700s other houses must have been built near by, for several families were said in the parish register to be 'of Town Head'. However, in the survey of 1716 they were listed as residents of the 'Top oth Towne'. These included the Shaws, Armitages and Blackburns. If we use the evidence of the survey to define the top of the town it seems that it may have consisted of all the housing to the west of the church. It certainly included '2 Cottages in the Market Place'. By 1778 the housing on this former area of waste had become a network of alleys and streets, and at least four public houses were located there. It is true that Huddersfield people still talk of the 'top of the town' but the term is now less precise.

TOWNING ROW

In the early 1800s, housing development at the west end of the town was stimulated by the building of new roads to Rochdale and Halifax. In both cases their starting point was the top end of Westgate, although the sections closest to the town would eventually be named Trinity Street and New North Road. Much of the development was of the villa type, but Towning Row was a more modest street, located behind what was then West Parade. In modern terms that is close to the junction of Greenhead Road and Trinity Street. I am puzzled by the name, which seems unlikely to be linked in any way with the town ings. It is more probably a surname and yet I have no evidence of a family called Towning in Huddersfield. Just possibly it was a variant of either Townend or Downing, surnames that were present in the town at that time. In 1880, when a footpath leading from New North Road was under debate in Council meetings, the street was actually called Townend Row. Much of the area is now occupied by the Technical College.

TOWSER

In Old English the word 'to touse' meant to shake about or to pull to pieces, and it was responsible for the name Towser, given to fierce dogs. This is on record from the 1600s and is said to have been the name in Huddersfield of a savage animal kept by the town constable. According to John Hanson the dog came to be associated with both the constable and his lock-up, which came to be known as Towser Castle. That may or may not be true but it should be said that more than one Yorkshire town had a prison called a Towser. The name is shown on several Huddersfield maps for the period 1818-26, located on what was then Castlegate.

T.P. WOODS (Gledholt)

On the 1791 map of Marsh these woods were called Gledholt Woods but they came to be known popularly as T.P. Woods when Thomas Pearson Crosland lived at Gledholt Hall. He was one of the original Improvement Commissioners in 1848, a J.P. in 1852, and he won the election of 1865 as a Liberal-Conservative. He is said to have been a humorist and impromptu speaker and was known to his friends as T.P.

TRINITY HOUSE

See West Field.

TRINITY STREET

In 1816, Benjamin Haigh Allen of Greenhead endowed the building of a new Anglican church in the town. It was called Holy Trinity Church and was completed in 1819. It is accessed from Wentworth Street via a lych-gate or from the street to the south via a handsome gateway. In 1820 the Lighting Committee authorised the placing of lamps 'as far as Trinity Church Gates' and it was not long before the name Trinity Street featured in the records.

TUNNEL STREET

See Springwood Street.

TURNBRIDGE

The district known as Turnbridge takes its name from the bridge over the canal at that point, erected to provide access to the area between the canal and the river. It was built as a swing bridge, so that loaded barges might pass, and is so described on the map of 1778. The present name is on record from the early 1800s although the structure has been altered several times in the course of its history. That part of the town may now seem something of a backwater but a letter to the Commissioners, in 1860, paints a quite different picture. The writer was Richard Armitage of Mountjoy House and he spoke of the 'turnbridge' as a nuisance, claiming that it was 'old, narrow and now nearly rotten', although it had been rebuilt only thirteen years previously. One section of the letter is worth quoting in full:

Looking down Trinity Street c.1930: Trinity House is to the left and Park Drive branches off to the right. There is a solitary tramcar in sight.
Clifford Stephenson collection

Turnbridge in 1849, with the early swing bridge leading onto Quay Street. The dock is shown at the bottom of Dock Street. Clifford Stephenson collection

... a fine wide street like Quay Street should no longer remain nearly a 'Cul de Sac' but should have a sufficient vent for the very greatly increased business, population and traffic in that direction – Quay Street itself is above 60½ feet in width and the present bridge is under 8½ feet wide. Both foot passengers and carriages are continually stopped or delayed, not only during the passage of boats, but also while each cart or carriage approaches and gets drawn over the bridge. A waggon today, with a heavy load on from my works, passed over it with great difficulty, and in endeavouring to guide the horses over the narrow track, our man was comparatively within a hair's breadth of being crushed to death. As it was the bridge got damaged ... one horse was thrown down (and) the waggon was fast on the bridge for an hour or two.

Turnbridge was at that time a busy, industrial place but it became also a residential area, principally as a result of the Artisans' Dwelling Act of 1874. This gave councils the right to erect houses, and Huddersfield Corporation responded very quickly, putting up several terraces at Turnbridge in the early 1880s. At a meeting in September 1881 it was resolved that Turnbridge Street should be 'sewered, paved, flagged and completed as a first class street'; it was reported also that twenty of the artisans' dwellings were already fit for occupation. They were among the first council-owned homes in the country and the streets were subsequently given such optimistic names as Violet, Hope, Rose, Lily and Daisy. However, they were to last less than a century and were demolished in 1970-71.

UNION BANK YARD (New Street)
The yard is named after the Halifax and Huddersfield Union Bank whose premises in 1837 were at 31 New Street. On the OS map of 1851 the bank fronted onto New Street, immediately to the south of the entry into the yard. The fluctuating history of the yard was well told in The Old Yards of Huddersfield and brought up to date more recently in The Buildings of Huddersfield. The bank moved in 1868 to the site now occupied by Lloyds and just a decade later the tenants of the buildings in the yard were mostly woollen merchants and manufacturers. The two exceptions were Ben Stocks, an architect, and Elliott Graham, a butter merchant. There are still good quality buildings in the yard which many Huddersfield people will remember with nostalgia as the 'home' of the 'Furniture Exchange'.

UNION STREET
Union Street formerly linked Northgate and Lowerhead Row and if this was part of its intended function it may explain the name. It was in existence by 1825 when it was resolved to fix a lamp there, and it is shown on the map of 1826 as partly developed, with a field called Tenter Crofts immediately to the south. The town's pinfold was located just beyond a range of buildings to the north, at the side of a path that passed over the canal by a footbridge. Just beyond the canal, in 1826, was the 'Gass House' which had opened six years earlier and was already supplying gas to the town's new lamps.

UNNA WAY (Ring Road)
The Ring Road underpass takes its name from Huddersfield's twin town in Germany. Unna is on the eastern edge of the industrial Ruhr.

UPPER GEORGE STREET
This is one of the two surviving stretches of George Street, and it is dealt with under that heading.

UPPERHEAD ROW
In 1815 'the Upperhead Row' was described as 'the street from Outcoat bank to the top of the town of Huddersfield'. In a lease of 1808 the tenant was held responsible for the maintenance of the street in front of his house - required to 'pave with Boulder or other Stones, and repair and maintain such parts of the street called Upper Head Row ... as is contained from the Crown or centre thereof'. See Lowerhead Row.

VAGRANT OFFICE (Fox Street)
The old Vagrant Office was located close to the buildings at Croft Head, directly behind the Cloth Hall. Little has been written about it, but it is shown on the OS map of 1851 and we can see from an early photograph that it was an unusual, two-storied building. The south end, to the side of the main entrance, has an elliptical façade and is built of stone. This may have been where the vagrant officer lived. The main part of the building is stuccoed and painted white; it has sash windows and there are flowers in a window box on the ground floor; a rooster is silhouetted against the stonework and three people are posed in front of the door. These are likely to be the vagrant officer (a Mr Jonathan Frost in 1853), his wife and daughter; the girl is holding a hoop in front of her. The minutes for 1868 contain a proposal to demolish the buildings and 'open out a street'. This was done the following year. At the same meeting they talked of 'enclosing and planting an open space at the foot of George Street'. This seems likely to have been the origin of 'Sparrow Park', formerly at the junction of George Street and Springwood Street.

VANCE'S BUILDINGS (Cloth Hall Street)
Vance's Buildings were on the south side of Cloth Hall Street adjoining the *King's Head Inn*, and the name is on record from 1845. The yard or court behind the buildings is shown on the OS map of 1851 and the directory of 1853 lists over thirty manufacturers who had premises there. For the most part they came from villages to the east and south-east, many of them from Kirkheaton parish and the Holme Valley, and two or three from Denby Dale and Penistone. The name Vance's Chambers can still be seen on the front of the building. Vance seems to have been another Scottish name.

VENN STREET
Henry Venn was possibly the most influential vicar in the town's history. He was appointed in 1759 and is credited by Edward Royle with having changed the face of religion in Huddersfield. He was obliged to return south in 1771

because of ill health and, although he was not forgotten by his parishioners, it was after his departure that the Church of England began to lose ground to the Dissenters and Methodists. The street that bears his name was not built until long afterwards although the junction with Kirkgate marks the site of the former vicarage. This was in the occupation of Henry Brook in 1849; it was used as a warehouse and was at the heart of an area described as 'ill contrived, unhealthy and disreputable' by Alexander Hathorne, the Ramsdens' local agent. He linked the property with the *Rose and Crown* and the *Fleece*, and suggested that 'a street could be formed right through the centre of it, extending from King Street to Kirkgate'. Despite that, it was to be over thirty years before the *Rose and Crown* was demolished and the building of Venn Street took place.

VERNON AVENUE
This is likely to be another example of a street name derived from a villa, for Joseph Batley was living at Vernon House on New North Road in 1879. The Vernon family of Haddon Hall in Derbyshire is likely to be the inspiration for the name, although the connection may be indirect.

VIADUCT STREET
The viaduct had been completed by 1848 when the Lighting Committee had a lamp placed under the archway 'leading from Bradley Spout to Bath Buildings'. That was on the line of the 'intended' John William Street. There were soon complaints about water seeping from the arches and it was proposed as early as 1861 that some of them should be enclosed. The paving of the street took place in 1850 and when it was finished it ran to the south of the viaduct, all the way from John William Street to Great Northern Street. A section of it was demolished when the Ring Road was built, thus isolating what is now Lower Viaduct Street. In the 1860s two companies had plans for new railway lines into Huddersfield and a second station in the Oxford Street area. The idea was to link this to the main station by a second viaduct but nothing came of the proposal.

VICTORIA BUILDINGS, VICTORIA LANE, VICTORIA STREET
Queen Victoria owed her name to her mother, the former Princess Victoire of Saxe-Coburg but she was christened Alexandrina Victoria in tribute to her godfather, Tsar Alexander I of Russia. Inevitably, numerous streets and buildings were named after her during her first years on the throne. These include Victoria Buildings, the first on record in 1842, Victoria Lane and Victoria Street (1849). The early evidence for the street names can be confusing, so I will deal with them separately. Victoria Lane presents fewer difficulties since it survives, running from King Street through to Ramsden Street. In 1849 it terminated at what is now Princess Alexandra Walk. Victoria Buildings was the property immediately to the west and in 1879 there were just two occupants, a woollen merchant called Sykes and a wholesale tea dealer called John Tetley. Victoria Street started from in front of Victoria Buildings, at right angles to Victoria Lane, and ran through to Queen Street. In 1879 the *Victoria Inn* was at no.8, with Jonas Horsfall as the landlord: he was also the secretary of the Victoria Loan Society. All that now survives of Victoria Street is Princess

The Viaduct Works on Viaduct Street (1877), seen through the arch near Alder Street. G Redmonds

Alexandra Walk. Perhaps the conflicting evidence of the town maps for c.1850 stems from the Trustees' doubts about the suitability of the name. The minutes preserve a suggestion that Victoria Lane should be re-named Shambles Lane, in order 'to discontinue the unworthy association of the name Victoria with a very minor street'. See Albert Hotel and Pig Market.

VULCAN INN (St Peter's Street), VULCAN STREET (Chapel Hill)
Vulcan was the Roman god who forged thunderbolts, and his name is often used in connection with furnaces and forges. For example, Vulcan Street off Chapel

Hill was the site of the Central Iron Works in 1905. There is a reference in the Commissioners' minutes for 1853 to a public house named the *Vulcan*, but it is not listed in White's directory that year and its location is uncertain. However, in 1879 the *Vulcan* was at no. 32 St Peter's Street, with George Walker as the landlord, and that is the site of the present inn.

WAKEFIELD ROAD

The meaning here is not in doubt, although the name emerged only after 1759 when the Wakefield to Austerlands turnpike was built. Previously 'road' was a relatively unusual word and the routes from one town to another were more often referred to as highways. The first Wakefield Road crossed the Colne at Aspley but then turned up Almondbury Bank at Mold Green, passing through Almondbury and Fenay Bridge, not Waterloo. The latter name is evidence of when the stretch of road through Dalton Greenside was constructed.

WAPPY NICK (Market Place)

This name is probably best known in connection with a locality on Lindley Moor, close to the present *Wappy Spring* public house. However, it is also an unofficial name for the alley called Market Walk that leads from the Market Place through to King Street. I made an appeal some years ago for information about this use of Wappy Nick and had several fascinating contributions. The late Mary Freeman explained to me that because the alley was so narrow it was necessary on busy Saturdays to push one's way through the throng. She demonstrated for me the dialect use of the verb 'to wap', meaning something like 'bustle' or 'jostle', but allied to a certain jauntiness. An alternative story is that 'wap' means home-brewed ale and that Wappy Nick was so named because a public house close by sold Wappy Stout, from the brewery on Lindley Moor. Whatever the meaning, there is no doubt that the name entered into local folk lore. One correspondent remembered his mother saying that 'only clever dicks ... could walk six abreast down Wappy Nick'.

WARRENFIELD HOUSE, WARRENSIDE (Deighton)

In the court roll of 1756, reference was made to a field in Fartown called the Warren, and the inference is that this had formerly been close to, or part of, an area reserved for the hunting of small game. If so the original connection was probably with Deighton Hall. The mansion house called Warrenfield is shown on the OS map of 1851, and Henry Brooke was living there in 1879. Warrenside is part of a recent housing development but may also owe its name to the Deighton warren.

WASP NEST ROAD (Fartown)

This road is part of an ancient highway that ran between Fartown Green and Longroyd Bridge, providing direct access to the upper end of the Colne Valley for travellers from places such as Bradford and Leeds. On the north side of the road, close to the junction with Halifax Old Road, there was formerly a public house or beer house called the *Wasp Nest*, although little is known about it. It is shown on the OS map of 1851 and Mr Joseph Wilkinson was the tenant in

1879. There have been other places with this name in the West Riding so the meaning should probably be taken at face value.

WATERGATE

This was another of the 'gate' names coined in the early 19th century. In what amounts to a relatively short history it has been applied to three different streets. On maps of the early 1800s it was part of the route that took people from Kirkgate and Rosemary Lane to Turnbridge - literally the gate or road to the 'water'. My first thought was that it must have referred to the water of the canal, but the river Colne was known locally as 'the water', so Watergate may have been thought of as the road to the river. That section later became Quay Street, but Watergate was preserved as a name and transferred to a narrow street to the south, beyond Dock Street. When that too was demolished the name was transferred again, taking the place of Castlegate, or at least what remained of it.

WATER LANE

This may have originated as a derogatory and unofficial name, for it was a narrow, slightly angled alley that linked Manchester Street and Albion Street: it lay roughly parallel to High Street, just to the south. It was listed as a street in 1837 and is clearly shown on the OS map of 1851 but it was in the area later developed as the Civic Centre and no trace of it has survived.

WATER ROYDS (Aspley)

Most recently this name was given to buildings in St Andrew's Road although it had originally referred to fields in the flat part of the valley east of Somerset Bridge. The suffix 'royd' implies that the area had been cleared in the 12th or 13th centuries and the fields may have been called the 'water' royds because they were under water at certain times of the year. It was considered good husbandry locally to enrich the waterside meadows by flooding them in winter and the practice involved making channels and sluices. These may have affected the course of the river: a comparison of the 18th century estate maps for this area reveals marked differences in the line of the Colne and it is possible that this was a man-made alteration rather than an act of Nature. As a result some pieces of land that were formerly in Dalton are now on the Huddersfield side of the river.

WATER STREET

In 1827, a special Act of Parliament empowered a number of named Commissioners to supply Huddersfield with water. They responded immediately, building the fashionable Waterworks Offices at the head of Spring Street. A tablet on the recently refurbished façade informs us that they were completed in 1828. Work was also carried out on the small reservoir behind the building and when the main reservoir at Longwood was finished, in 1829, water was piped down from it to this service tank. Water Street, like Spring Street, has some fine town houses but I am uncertain just when they were built. The street is first mentioned by name in the Commissioners' minutes of 1865.

Waverley House, Wentworth Street, at the junction with New North Road. Dr William Scott lived here in 1879. G Redmonds

WAVERLEY HOUSE, WAVERLEY ROAD (Wentworth Street)

Waverley is a word with a curious history. It is found in Surrey as the name of a small hamlet - with a disputed etymology – and this gave rise to a rare surname that was on the verge of extinction in 1881. It is likely, therefore, that Waverley's popularity as a house and street name depends almost entirely on the prominence given to the surname by Sir Walter Scott. He chose Edward Waverley as the name of his young romantic hero and the wide-spread appeal of the so-called Waverley novels then inspired any number of minor place-names. Edinburgh's Waverley Bridge and Waverley Station are perhaps the best known. Waverley Road in Huddersfield was probably named after Waverley House which is located at the New North Road end of the street. In 1879 William Scott lived there, a doctor who had graduated in Edinburgh, so it is easy to understand why he would choose Waverley as the name of his house. It was not a very important street, doing little more than provide access to Wentworth Street, and I have not located the name in 19th century sources. The United Reform Church was built there in 1910 and that may have improved the street's status. There is another Waverley House in Edgerton and the former Waverley Hotel was in Kirkgate.

WELL or WELLS (Beast Market)

The house known as the 'Well' was located just to the north of the Beast Market and no doubt took its name from a well on the site – possibly a town well. For at least two centuries it was the home of a family called Horsfall, best known perhaps for William Horsfall the victim of the Luddites. The parish register has an entry for the Horsfalls of Well in 1611 and John Horsfall of Well was a merchant in the 1820s. On the OS map of 1851 the house was called the 'Wells' and the building, if not the name, seems to have survived into the 20th century.

WENTWORTH STREET
The early references to this name in the Commissioners' minutes illustrate how difficult it can be to say exactly when a street was built. In 1861, for example, a dangerous well in Wentworth Street is mentioned, and yet it was another five years before Captain Graham gave the Commissioners notice of his 'intention to lay out the New Street ... to be called Wentworth Street'. It was a street for middle class families and in the 1879 directory all the residents but one were given the title of 'Mr' or 'Mrs'. The exception was William Platts, a joiner. Nevertheless, it was being claimed in 1867 that recent building had made it impossible for the scavengers' carts to empty 'the privies belonging to the houses in Wentworth Street'. The source of the name seems likely to be in the close relationship that the Ramsden family had with Earl Fitzwilliam of Wentworth.

WESTBOURNE ROAD (Marsh)
There are several localities called Westbourne in the south of England but it is not clear to me why the name should have become popular in some northern towns in the 19th century. In 1879, George Winter Rhodes lived at Westbourne House in Queen Street South and later there was a Westbourne Terrace in Marsh. This probably gave its name to the adjoining part of New Hey Road. Perhaps the London district of Westbourne Park was the inspiration. The 'bourne' or stream that is linked to this latter place-name is now visible only in Hyde Park where it is dammed and widened to form the Serpentine.

WEST FIELD (Trinity Street)
It can be very confusing trying to identify this name, for there were several places so called on Trinity Street in the 1800s. The most important was certainly the fine terrace of houses located just to the east of West Place; this lies in front of Holy Trinity Church and has been known more recently as Trinity House. These were fashionable residences and many of the earliest tenants were woollen merchants. The church was built in 1819 and West Field and West Place must date to roughly the same period, since all three are shown on the map of 1826. Together they form an elegant group of buildings, very evocative of Georgian Huddersfield. Also on Trinity Street were two other West Fields and a West Field Terrace, all shown on the OS map of 1851. The repetition of the name might suggest that West Field referred originally to one of the town's open fields but as yet I have no evidence for that.

WESTGATE
Westgate is the natural extension of the old town gate, or Kirkgate as it was later called, and, like Kirkgate, it was a name coined at the very end of the 18th century. The choice was logical, for it was the town's western exit, but the

impression we receive from estate documents is that Sir John Ramsden did not decide to give it that name immediately. For example, it is called West Street on the town map of 1797 and Westgate in the rental of 1798. Of course at that time the top of town was still a maze of alleys, and there were local routes to Marsh and Lindley. However, there was no Trinity Street and no New North Road, and travellers to Halifax still left the town via Northgate.

WEST PARADE

West Parade was formerly the name given to the section of Trinity Street that ran between New North Road and Greenhead Road, and much of it was demolished when the Ring Road was built. Records show that it had replaced the name Greenside by 1834 and it was clearly a more fashionable address. In 1837, though, the residents of West Parade represented a wide cross-section of the working population and included cloth dressers, wool staplers, a carrier, a brick-maker and a surgeon. The entry I prefer is the one for John Brunton who lived at No. 67; he operated a beer-house and supplemented his income as an 'artificial leg, arm, hand, and spring truss maker'.

WEST PLACE (Trinity Street)

In the context of street names the word 'place' usually refers to an open space or square, as in the Market Place, but it was used also for an individual dwelling house or mansion. It is this meaning, I believe, that explains West Place, first found on the map of 1826. The name is seldom used now, possibly because the building is part of the property occupied by the Highfield Funeral Service, but in the past the house was a landmark on the road to Marsh.

WHARF INN (Aspley)

There has been a public house called the *Wharf* on this site for at least two centuries, although the present building cannot be very old. It takes its name from the canal wharf and may therefore date from when the canal was built. The inn is shown on all the early 19th century maps but the first documentary reference to the name is in the Brewsters' Sessions for 1806, when John Senior was the landlord. It has recently been re-named the *New Wharf*.

WHEAT HOUSE, WHEAT HOUSE ROAD (Birkby)

I have found no early references to Wheat House, and I am reluctant to link it with other Fartown 'houses' such as Black House, Flash House and Green House. All those hamlets were established by the 16th century, whereas Wheat House appears to date only from the early 1800s. In 1837 it was the home of a fancy goods manufacturer called John Beaumont, a house with extensive grounds located on the west side of Blacker Road. It is possible that it was a villa name but if that is the case the element 'wheat' is unusual. It later gave its name to the new road that linked Bay Hall and Birkby Hall.

WHITACRE LODGE (Leeds Road), WHITACRE STREET (Deighton)

Professor Smith thought that Whitacre Lodge might be the source of the surname Whitaker, and he tentatively identified it with place-names recorded in 13th century sources. In fact these refer to a locality that lies outside the township of Huddersfield, and Whitacre Lodge has a quite different origin. It was the name of a villa close to Leeds Road and may have been erected for the mill-owner and merchant John Whitacre. He was a friend of Richard Oastler and responsible for the building of Christ Church in 1823-24. Whitacre Street in Deighton led down to the river where Mr Whitacre's mill was located.

WHITE BEAR, WHITE BOAR (Market Place)

In one or two early references to this public house, the reading of the name could be the *White Boar*, but it was certainly the *White Bear* in the early 19th century. The location and the fact that it had a succession of landlords called Walker confirm the identification. Both names have long histories in England and both are heraldic in origin. The boar was an allusion to Richard III whilst the bear was an emblem of the Earls of Kent. Unusually the public house is mentioned by name in the court roll of 1734 when John Walker was indicted for short measure offences. The last known landlord was Joseph Durrans who held the licence from 1812 to 1828.

WHITE CROSS (Bradley)

The probability is that there was once a pole or stone pillar here, marking the cross roads, and that it was painted white to attract attention. These crosses were common features in the neighbourhood but they had no religious significance that I know of. There were others at Marsh, Mold Green, Cowcliffe and Sheepridge. The public house goes back to 1803 at least, when the landlord John Hubank was referred to in the Brewster Sessions. When the Bradley estate was sold, in 1829, it was described as a 'well accustomed Public House near Cooper Bridge' with stabling, outbuildings and over thirty-six acres of land. John Howgate was then in occupation.

WHITE HART (Cloth Hall Street)

The white hart was a heraldic symbol associated with Richard II, and the earliest inn to be so named was in his reign. In Huddersfield there was a *White Hart* in 1777 but it was in the Croft Head area and has no known links with the more famous hostelry that now stands opposite Sainsbury's. Sainsbury's replaced the cinema that had been built on the site of the Cloth Hall and an evocative picture of c.1860 shows the *White Hart* facing the Cloth Hall – in closer proximity than they were in reality. It was a popular venue for clothiers at that time so perhaps the artist intended to show us how closely linked the two places were commercially. Edward Law's history of the *White Hart* goes back to 1791 when the landlord was the aptly named John Tavenor. When it was put up for sale in 1821 it was advertised as having stabling for fifty horses and five acres of grassland close to the town. During renovations to the façade of the inn, in 1981, the stucco was removed and it was possible to see the locations of the windows which had been filled in over a century earlier.

WHITE HORSE (Beast Market)

This was probably one of Huddersfield's oldest hostelries, identified by Edward

Law as the 'house' of William Kay, innkeeper, in 1747-49. Other families known to have held the tenancy were the Booths and Clays, and Sarah Clay was the landlady in 1857 when the inn was demolished. It was removed so that Lord Street could be opened out, as part of the New Town development.

WHITE LION
This is another popular name for a public house that derives from heraldry, often with royal connections. It has a long history and has been noted as early as 1512. In Huddersfield the name apparently occurs twice in 18th century records but the *White Lion* in Cross Church Street was the better known establishment. A family called Langley was connected with the inn for nearly thirty years, starting with William Langley who was the landlord in 1797.

WIGGAN LANE (Sheepridge)
This was an ancient right of way that diverged from the main highway to Leeds at the top of Sheepridge, en route to Rastrick and Bradford. Significantly, part of it is still called Old Lane but the first evidence I have for the name Wiggan Lane is the OS map of 1854, and the origin remains uncertain. It appears to derive from Wiggin House in Sheepridge, the home of a solicitor called John Clay in 1853 but, even if that is the case, we are still no nearer to an explanation of 'Wiggan'. One possibility is that it derives from the surname Wiggan, found in Yorkshire from an early date. In that case the ultimate origin might be either the Lancashire place-name Wigan or a personal name brought here at the time of the Norman Conquest. A typical example is Wygan le Breton (1252).

WILKS YARD
See Hammonds Yard.

WILLOW LANE (Fartown)
There would formerly have been willows here, by the town brook, so the name can probably be taken at face value. I have included it here because the lane may have a longer history than I first imagined. Certainly its role has changed since Bradford Road and St John's Church were built. The name is shown on the OS map of 1851 and the lane is shown without name on the map of 1780, where it appears to run south-east from Bay Hall, along the side of a field called Mag Croft. I find no evidence for it on the map of 1716, although Magg Croft is there.

WOODHOUSE (Fartown)
There are several hamlets called Woodhouse in the Huddersfield area, all associated with major 'villages' and all located at some distance from the village centres. Just when they originated is uncertain but the earliest references occur in the two centuries that followed the Norman Conquest. As the population increased so did the need for food, and the 'Woodhouses' are proof that it became necessary to make further woodland clearances and bring more hard-won land under the plough. These new settlements were typically called Woodhouse or just Wood, and there are classic examples at Shelley

Woodhouse, Emley Woodhouse and Wood in Fixby. The Fartown hamlet called Woodhouse is mentioned in a land conveyance of 1383 but it almost certainly had its origins much earlier, well before the Black Death. We know something of its tenurial history and can identify it, from 1452 at least, as a farm in the occupation of yet another branch of the Brook family. By the 1630s there were at least two 'messuages' at Woodhouse, and other named tenants include John Horsfall, William Jepson and Arthur Swallow. From 1786 to c.1844 the estate was held by the prosperous Whitacre family, and they were responsible for the building of Christ Church in 1823-24. It is said that old dwelling houses were demolished to make way for the new church which may therefore mark the medieval settlement site. See also Whitacre Lodge.

WOODLAND MOUNT (Trinity Street)
As the urbanisation of the Huddersfield landscape increased, more and more of the new minor place-names were given 'wood' as an element. There was a Woodland Cottage, a Woodlands, a Woodlands House, a Woodland Mount and many more. The last-named is an attractive, ashlar-fronted terrace at the top end of Trinity Street, and the occupants in 1879 were men such as Mr James Oates Bairstow and Arthur Smith, an architect. The terrace may date from just before 1853 when the merchant Thomas Gledhill was said to be of Woodland Mount.

WOODMAN INN (Bradley)
In the Huddersfield area public houses with this name are usually found in locations close to former woodland estates. Bradley was such an estate, owned by the Pilkingtons for about 350 years, so the inn may have a much longer history than the evidence for the name suggests. It seems possible that successive woodmen supplemented their income by selling ale and beer. Ironically, though, the first reference to the place-name is in 1829 when the Pilkingtons sold the estate. It was described in the sale plan as a public house or inn, with John Gibson as the under tenant.

WOODSIDE, WOODSIDE LANE (Sheepridge)
Although Bradley was part of Huddersfield township, it had a quite separate history as a manor, and for centuries was held by the Pilkingtons. They sold it in 1829 and the sales details make it clear that the woods were then still surrounded by a ring fence. It is a boundary that can still be discerned from the Bradford Road, on the line of what is now aptly called Woodside Lane. Close by there was formerly a farm called Woodside, dating from the early 1600s; it was the residence for several generations of a family called Cowper or Cooper - possibly kinsmen of the Coopers of Edgerton.

WOODTHORPE TERRACE (Longroyd Bridge)
Woodthorpe Terrace is prominently located on the hillside above Fenton Square and Spring Lodge. It takes its name from a villa called Woodthorpe House which may have been built in the early 1800s. There are precipitous flights of steps up this hillside and in 1856 they proved too much for the mill-owner

John Starkey who collapsed as he was returning home to Springwood. He died shortly afterwards, aged sixty-five.

WOOLCROFT
Few Huddersfield people will have heard of this fascinating field name and yet in many ways it highlights the dramatic changes that have taken place in the urban area over the centuries. In two 13th century documents the name is spelt Wlfcroft and Wolvecroft and these prove that the first element was 'wolf' not 'wool'. This may have been a personal name but I prefer to see it as a reference to the animal. The early maps show that the field was located between the top end of the town and the hamlet of New House, somewhere in the area where St Patrick's Church and the former Infirmary are located. It is difficult now to see this as a clearance, with wolves in the surrounding wood, but local place-names such as Wooldale and Woolrow (Shelley) support the origin.

WOOLPACK (New Street)
The *Woolpack* in Huddersfield stood at the top end of Ramsden Street, on the south side of the junction with what was then Buxton Road. John Hanson could remember it from when he was 'but a little tiny boy' and it is mentioned in the Brewster sessions of 1803. Edward Law noted that Joseph Kaye, who was Huddersfield's most famous builder, lived there for a time in the early 1800s. The vernacular style building was closed as an inn in 1928 and demolished in 1964-65, in the first phase of the town centre redevelopment. Not surprisingly the name is a popular one for public houses in Yorkshire, and many signs have an illustration of the traditional bale or pack of wool, said to have weighed 240 pounds.

WORMALD'S YARD
This is one of the recently refurbished King Street yards, given a new lease of life in the development associated with Kingsgate. Apparently there were still nine tenants living in the yard as recently as 1956 but it was falling into ruin when The Old Yards of Huddersfield was published thirty years later. The authors had also discovered that it was sometimes called Edward's Yard, notably in 1866 when a saddler called John Wormald was living at no. 80, King Street. This was just on the top side of the arched entrance.

YATE HOUSE (Lost)
This means the 'gate house', and the name featured in the records between 1524 and 1664. However, it had apparently fallen out of use by the 1700s and the location of the gate that gave its name to the farm is not known. The tenants were members of the Brook family throughout that period, closely linked to the Brooks of Bay Hall, so Yatehouse was almost certainly in Fartown. Indeed, there are references in the court rolls which show that the tenants were responsible for the maintenance of parts of Blackhouse Brook and the town lane. In 1655, several men were ordered to 'repair the way from the Feildyate to the Yatehowse', so the 'gate' was probably a significant location on the town lane.

YORK PLACE (New North Road)

This was not a 'place' in the sense of a square but the name of a short street or terrace. It was located along the east side of the Infirmary grounds and was a fashionable address in the part of the town favoured by successful tradesmen and professionals. The houses are not shown on the map of 1826 but may have been built in the early 1830s, soon after the completion of the Infirmary (1829-31). Among the residents listed in the directory of 1837 were S. & H. Haigh, milliners, William Wilby, a wine merchant, and Thomas Robinson, a solicitor. The terrace did not survive the building of the Ring Road and the expansion of the Technical College.

YORK STREET

Although York Street was part of the New Town development and had a fashionable name, it was never an important street. It lay between Union Street and Northumberland Street, and stretched from Northgate to Hawke Street. Nothing of it has survived but part of the land that it occupied lies behind the buildings east of the Ring Road. It is first recorded on the OS map of 1851 and the entry in the directory of 1853 makes it clear that it was still in the planning stages. Even in 1879 only three residents were listed, a shopkeeper, a coal dealer and a shoemaker.

ZETLAND STREET, THE ZETLAND HOTEL

The Earl of Zetland was the brother of Mrs Isabella Ramsden and he was one of the Trustees who administered the estate during her son's minority. It is clear therefore how Zetland came to be a Huddersfield place-name but less obvious perhaps that it is an alternative form of Shetland. The hotel came first, in 1847, and Zetland Street was named soon afterwards. It was called 'an intended new street' in 1850 and yet two years earlier the Commissioners had decided that 'vacant ground in Zetland Street' should be used as a depot for paving and drainage materials.

SOURCES

I have provided no footnotes in this book, and there are two reasons for that omission. I began with the intention of annotating everything carefully but abandoned the plan when it became clear to me that the sheer number of references might prove a real hindrance to the potential reader. I am conscious of the inconvenience this will cause to scholars but I am hoping that these very full comments will help to identify exactly where and how I came by the information. To that end I have tried to link the text to the sources by quoting significant dates, and names of authors wherever possible. In partial mitigation I should add that I have already published footnotes to many of the place-names and they can be found in the books and articles listed below, particularly those of more recent date.

Some of the evidence I have used is drawn from a variety of familiar sources, notably the various census returns, several local newspapers, Acts of Parliament, and the parish registers, including those for Elland and Almondbury. The Acts deserve special mention for they are much less familiar to local historians than perhaps they should be. Many of them related to special enterprises such as the Broad Canal (1774), the Narrow Canal (1794) and the Railway (1845-1872) and these brought about significant changes that are visible in the landscape and reflected in local records. The Acts that authorised new turnpike roads are complemented by the account books and associated documents of the trustees, and most of these records are located in Wakefield, classified under RT in the County Record Office. There were also Acts behind amenities such as the Waterworks (1827), the Burial Ground (1852) and the provision of Gas (1861), but perhaps even more important were the Acts that marked the stages of Huddersfield's independence and progress. These were the Lighting Act of 1820, a number of Improvement Acts from 1848 and the Corporation Act of 1882.

Not all the information in the book is drawn from readily identifiable sources. Over the years I have had conversations and correspondence with many knowledgeable acquaintances, friends and colleagues, and sometimes what they have spoken about or written has materially affected my own opinions. In some cases I have mentioned the person by name in the text but I acknowledge here the work done by Sandi Hewlett on John Dobson, the banker, and that of Michael B. Fisk on Bradley Hall. Moreover, I have frequently made use of information from gravestones, dated buildings, sketches, prints and old photographs, notably the series of photographs by Isaac Hordern in Huddersfield Library. Some of those items are in my own possession and others are not generally available but in such cases I hope I have made the source clear in the text. For example, I chanced to walk past the *White Hart* twenty-five years ago as workmen were removing the stucco that covered the front of the building. That allowed me to sketch the locations of the 18th-century windows before they were once again covered over.

Important evidence can also be apparent in the pattern of the streets or the style of the buildings. For example, the observant pedestrian is bound to notice the 'vernacular' aspects of the houses at the bottom end of Greenhead Road or of inns such as the *Commercial* and the *Plumbers*. The buildings in these cases provide us with information that is just as valid as a documentary reference. I have repeatedly walked through the streets of Huddersfield as part of the preparation for this book and seen much that I had previously missed.

Town Histories:
There are numerous published sources on Huddersfield and the town has been fortunate in the number and quality of its chroniclers. Philip Ahier did much useful work at a time when many sources were still not generally available, whilst both Stanley Chadwick and Edward Law have more recently published excellent books and articles. Many other works are of varying quality but even when the authors' history is not always trustworthy there is usually something that is useful, if only a comment on contemporary events. Hobkirk, for example, writing in 1868, began with a valuable description of the town as he saw it and there are similar essays in many of the directories.

A number of specialised publications also contain facts that contribute to our knowledge of Huddersfield's development. There are far too many for me to list them all here and I am omitting histories of institutions such as schools, churches, chapels, mills and cooperative stores, but they should not be forgotten by those seeking more information. There are also autobiographies or family histories which contain isolated but relevant items, and I would mention, as examples, Family Memorials of the Houghtons (1846) and Canon Bateman's Clerical Reminiscences (1882). G.S Phillips, in his Walks round Huddersfield (1848) was writing as the town was in the throes of change and makes several revealing comments. However, the sources which have been of most use are those listed below.

C.A. Hulbert, Annals of the Church and Parish of Almondbury (1882).
D.F.E. Sykes, Huddersfield and its Vicinity (1898). See in particular chapter XII.
T.W. Woodhead, History of the Huddersfield Water Supplies (1939).
P. Ahier, The Story of the Three Parish Churches of St Peter the Apostle, Huddersfield (1948).
W.B. Crump, Huddersfield Highways down the Ages (1949) Reprint 1968.
R. Brook, The Story of Huddersfield (1968). At pages 112 and 152 are copies of town centre maps for 1826 (George Crosland) and 1850 (Robert Nixon). The section of Jefferys' map for the Huddersfield area (1772) is at page 52.
C. Pearce (ed.), Urban Studies Series (Vols 1 & 2), 1977, 1978.
M.L. Faull and S. A. Moorhouse (eds), West Yorkshire: an Archaeological Survey to A.D. 1500 (1981). There is a good summary of Huddersfield's administrative and tenurial history in Vol. 2.
E.J. Law, 18th Century Huddersfield: the day books of John Turner (1985).
L. Browning and R.K. Senior, The Old Yards of Huddersfield (1986).
E.J. Law, Joseph Kaye: builder of Huddersfield c.1779 to 1858 (1989).
E.A.H. Haigh (ed.), Huddersfield: a most handsome town (1992). This is a fine collection of useful and scholarly essays. Of particular relevance here are those by G. Redmonds, E. J. Law, D. Whomsley, E. Royle, R. Dennis and J. Springett.

G. Redmonds, <u>The Making of Huddersfield</u> (2003).

K. Gibson and A. Booth, <u>The Buildings of Huddersfield: an illustrated architectural history</u> (2005).

Place-Names:

T. Dyson, <u>Place-Names and Surnames</u> (1944). Sadly this book cannot be recommended. Although it contains some items of interest, too many of the etymologies offered are inaccurate. The subject has moved on since it was written: methods have changed and much new material is now available. It does not concern itself with the names of the streets.

A.H. Smith, <u>The Place-Names of the West Riding of Yorkshire</u>, Part II, pp. 295-300 (1961). I have my reservations about this mammoth work as will be clear in the text.

G. Redmonds, <u>Old Huddersfield 1500-1800</u> (1981). This contains detailed source references to over sixty town centre place-names. Other place-names are dealt with in the article 'Settlement in Huddersfield before 1800', which is chapter 2 in <u>Huddersfield: a most handsome town.</u> See above.

Articles and pamphlets:

J. Burhouse and C. Pearce, 'Huddersfield in 1849: an eyewitness account', <u>Old West Riding,</u> Vol. 2(1).

D.L. Clarkson, 'An Outbreak of Cholera at Paddock in 1849', <u>Old West Riding</u>, Vol. 7((1)

D.L. Clarkson, 'St George's Square and the New Town of Huddersfield', <u>Old West Riding</u>, New Series, Vol 9.

E.J. Law, 'An Exemplary Barn', <u>Old West Riding</u>, Vol. 4(2).

E.J. Law, 'The Bradley family and their New House', <u>Old West Riding</u>, Vol 5(2).

G. Redmonds, 'Colne or Holme', <u>Old West Riding</u>, Vol.2(2).

D. Whomsley, 'A Directory of Huddersfield and District for 1791', <u>Old West Riding</u>, Vol. 3(2).

C. Thackray, 'An American View of Huddersfield in the 1870s', <u>Old West Riding</u>, New Series, Vol. 15.

G. Redmonds, 'Personal names and surnames in some West Yorkshire royds', <u>Nomina,</u> Vol. 9 (1985).

H. Heaton, 'Yorkshire Cloth Traders in the United States 1770-1840', <u>Miscellany</u>, Thoresby Society, Vol. XXXVII, pp. 225-287.

Anon. 'College Recollections', <u>Huddersfield College Magazine,</u> No. 6, Vol. VI and No. 10, Vol. VI (1878). These are little-known but fascinating reminiscences of the College and the area around Highfields. They relate to just before and after the school was built.

P. Ahier, Huddersfield and its Manors, Cuttings Book.

Directories:

Huddersfield Local History Library holds numerous trade directories, dating back to Baines's of 1822, but for earlier sources the most accessible copies are those in Edward J. Law's <u>Essays in Local History</u>, Nos 1-4. There he covers the years 1732 to 1817. D. Whomsley wrote about and transcribed a little-known directory of Huddersfield and District for the year 1791 in the journal Old West Riding. See above.

Reminiscences:
An extremely valuable series of articles was published in the Huddersfield Examiner in 1878, between 25 May and 8 June. The series recalls the years about the turn of the century and was written by John Hanson who signed himself 'Native'. Also of interest is a long article by D. Schofield published on 10 September, 1883. It touched on the years 1825-26.

Documentary:
Much of the documentation consulted for place-name evidence had its origins in the administration of Huddersfield and its neighbours as landed estates. Little is known about some of these connections but they should not be ignored. For example, occasional facts can be gleaned from the Gascoignes' possession of parts of Kirkheaton in the Middle Ages and the later holdings of the Lister Kayes in places such as Marsh. A few references have been made to deeds in the Spencer/Stanhope estate papers in Bradford and the Armytage papers in Halifax, both of which are repositories within the WYAS. Developments on the lands belonging to major freeholders can also pose problems, but I have found good information in material relating to the Fentons (see below) and valuable maps were drawn for Marmaduke Hebden and Hirst and Kennet (see Maps). The following are housed in Huddersfield Library as part of Kirklees Archives, West Yorkshire Archive Service (WYAS).
1) The Ramsden family estate records. These include title deeds (DD/R/dd/1-8), account books (DD/RA/f), estate surveys and plans (DD/RE/s) and rentals (DD/RE/r). Very important are the court rolls of Almondbury, 1627-91 (DD/R/m). Despite the title the material covers Huddersfield as well. The estate correspondence during the trusteeship of 1839-53 (DD/RE/C) emphasises the role of the agents and Mrs Isabella Ramsden at a watershed in the town's growth. Among these papers are numerous small maps and plans. One 'stray' Ramsden document of real significance is a tripartite marriage settlement of 1624 (RA/51), surviving in the family's estate papers in Sheepscar, Leeds (WYAS).
2) The Whitley Beaumont estate records. These are not primarily concerned with Huddersfield, but the Beaumont family held half the lordship of Huddersfield until the 17th century and there are many deeds in particular which provide vital evidence for the earlier centuries. These are classified as DD/WBD/I (combined properties) and DD/WBD/VIII (Huddersfield). An isolated fragment of court roll for 1532-33 (WBR/2), although difficult to read, contains numerous items of interest and makes us rue the loss of the remainder of the series.
3) The Clarke-Thornhill estate records. These are not principally concerned with Huddersfield but the records relating to Fixby can be useful. Many of the deeds have been published by the Yorkshire Archaeological Society in their Record Series but the collection merits further attention. It is in the keeping of the Yorkshire Archaeological Society in Leeds, housed at their Claremont

headquarters and classified under DD/12. Some categories of deeds (T/DD), leases (T/L) and surveys (T/S) are in Kirklees Archives.

4) The Philip Ahier MSS. These are classified under DD/AH and contain the Fartown Surveyors' Account Book, 1715-1790.

5) J.W. Wilson and Son, tinsmiths, Lockwood, records 1800-1920 including diaries 1820-1920 (B/JWW).

6) G.W. Tomlinson collection of materials relating to Huddersfield (KC174).

7) The minutes of the Commissioners and then the Town Council throw a great deal of light on the history of the streets and those that I have read in the original are listed below. I have not been through all the books so there is certainly scope for more work on the subject. There are also published volumes of Council Minutes in the Library and I have occasionally quoted from those, for example the records of 1880-81 and 1894-96.

The Minutes of the Lighting, Watching and Cleansing Committee, 1820-43 (KHT 2/1).

The Minutes of the Commissioners, Book B, 1843-48 (KHT 2/2).

The Minutes of the Improvement Commissioners, Book A, 1848-1859 (KHT 9/2/1).

The Minutes of the Paving Committee, 1848-53 (KHT 9/8/1): Ibid. 1853-56 (KHT 9/8/2)

The Minutes of the General Purpose Committee, 1848-56 (KHT 9/16/1).

The Minutes of the Drainage Committee, 1852-62 (KHT 9/7/2).

The Minutes of the Paving, Drainage and Works Committee, 1866-68 and 1868-73 (KHT 9/9/2).

8) Unfortunately, Huddersfield, unlike many neighbouring towns and cities, lacks published volumes of original wills. Many of those I have referred to are transcribed copies that are in my own possession, notably those for Bay Hall. This is one relatively unexploited source that might provide significant new evidence.

9) Huddersfield enclosure award (no map) 1789 E/H.

10) The Quarter Sessions Records of the West Riding are held in the County Record Office in Wakefield and a full guide has been published (1984). Some information used in this book has to do with roads and bridges and can be found via the indexed Indictment Books (QS4) and Order Books (QS10). Land tax records for 1752-1832 are in QE13 and licensed houses for 1697-1828 in QE32. Occasional items were located in the Rolls (QS1).

11) The solicitors Eaton, Smith and Downey formerly had numerous clients in Huddersfield. Their records are in the County Record Office in Wakefield, classified as C296, and they contain the map and accounts of Lewis Fenton's estate. I should add that in 1857 the estate was being managed by trustees, as I have elsewhere implied that Lewis Fenton was still alive at that time.

12) Some important and relevant primary sources are in print, having already been transcribed and/or translated. They include the poll tax of 1379, subsidy rolls of 1524, 1545 and 1588, and the hearth tax of 1672. For other evidence, some of it much earlier, there are deeds published in the Yorkshire Archaeological Society's Journal or Record Series and W. Farrer's Early Yorkshire Charters (1916).

13) Two extremely interesting trials throw light on Huddersfield place-names and aspects of the town's development. The first is Ramsden v Thornton, the tenant-right case of 1866, with numerous full and detailed depositions. The second is Middlemost Brothers & Co. v The Borough of Huddersfield (1905) chiefly of value for the Birkby area and Norman Park. Transcripts of both are held in Huddersfield Public Library.

A note on Bradley
Many of the above records cover both Bradley and Huddersfield but Bradley's separate role as a grange of Fountains Abbey and then a woodland estate under the Pilkingtons, means that it has alternative, independent sources. For the early period the two volumes of W.T. Lancaster's Chartulary of Fountains Abbey (1915) are essential and for the period up to 1829 there is much of interest in the Pilkington MSS. The Spencer Stanhope papers in Bradford Archives hold deeds, correspondence etc. that is particularly relevant to the area around Colne Bridge, and the history of the forge there is in G. Redmonds's The Heirs of Woodsome (1982).

Maps
The documentary evidence must of course be related to maps and to the historic landscape. The following list of sources includes several which contain one or two items only but these are often of real interest. As far as I am aware there are copies in Huddersfield, either in the Local History Library or in the Archives. One or two are likely to be paper tracings or copies that I have redrawn and deposited there myself. An excellent article by E.J. Law rightly draws attention to problems raised by the estate maps.
1634 Survey of the Ramsden Estate in Almondbury: W. Senior.
1716 A Mapp of The Estate Belonging to ... William Ramsden In ... Huddersfield: T. Oldfield.
1757 Plan of the River Calder: John Smeaton.
1768 A Map of an Estate belonging to Mr Hebden and others ... .T. Hogg.
1773 A Plan of the intended Navigable Canal from Cooper Bridge to Huddersfield: Faden and Jefferys.
1776 Plan of the Canal now making: Nicholas Brown.
1778 Plan of the Town of Huddersfield ... belonging to Sir John Ramsden: W. Whitelock.
c.1780 A Map of the Estate belonging to Hirst and Kennet, Esqrs. I came across this undated map in John Goodchild's collection in 1981, shortly after completing Old Huddersfield, and had time only to summarise its content in a note on p. 55. I suggested a date of c.1775 but find Edward Law's arguments for a date between 1779 and 1782 convincing.
1780 Plan of Huddersfield Estate in connection with a book of reference.
1791 A Plan of an Estate situate at Marsh in Huddersfield ... belonging to John Lister Kaye: Jonathon Teal.
1797 This estate map is linked to the estate valuation of 1797 but Edward Law makes a good case for placing it before 1780.
1818 Plan of the Town of Huddersfield.

1820 A Plan of the Town of Huddersfield.

1825 Plan of the Town of Huddersfield: T. Dinsley.

1826 A Plan of Huddersfield: George Crosland.

1829 The Manor or Lordship of Bradley: a sales plan.

1843 The 1 inch Ordnance Survey map of Huddersfield. Old Series, sheet 88. The topographical survey was started and finished in the three years preceding publication.

1851 The 24 inch Ordnance Survey map of Huddersfield, in 13 sheets. This was surveyed in 1848-49.

1854 The 6 inch Ordnance Survey map of Huddersfield, sheet 246. This was surveyed in 1848-50.

Later OS maps are referred to in the text by date.

GEORGE REDMONDS - Publications

1	Yorkshire: West Riding, English Surnames Series, Vol. 1	Phillimore, 1973 (314pp)
2	Surnames around Huddersfield,	Huddersfield Examiner, 1974 (55pp)
3	Old Huddersfield: 1500-1800,	GR Books, Huddersfield 1981 (58pp)
4	The Heirs of Woodsome, and other essays,	GR Books, Huddersfield 1982 (64pp)
5	Photographs of Old Lepton,	GR Books, Huddersfield 1983 (26pp)
6	Almondbury: Places and Place-names,	GR Books, Huddersfield 1983 (63pp)
7	Huddersfield and District under the Stuarts,	GR Books, Huddersfield 1985 (54pp)
8	Changing Huddersfield,	Kirklees Libraries, 1985 (79pp) Photographs
9	Ed. 'David Bower M.P.' Old West Riding, Vol. 7 (2)	Old West Riding Books, 1987 (32pp)
10	Slaithwaite:Places and Place-names,	GR Books, Huddersfield 1988 (52pp)
11	Yorkshire Surnames Series, Part I: Bradford and District,	GR Books, Huddersfield 1990 (62pp)
12	Yorkshire Surnames Series, Part II: Huddersfield and District,	GR Books, Huddersfield 1992 (64pp)
13	Holmfirth: Place-names and Settlement,	GR Books, Huddersfield 1994 (70pp)
14	Surnames and Genealogy: a new approach,	New England Historic Genealogical Society, Boston, U.S.A. 1997 (310pp)
15	Yorkshire Deeds in Kansas,	GR Books, Huddersfield 2000 (154pp)
16	Yorkshire Surnames Series, Part III: Halifax and District,	D. Shore, Huddersfield 2001 (104pp) Illustrated
17	Yorkshire Surnames and the Hearth Tax Returns of 1672-73 (with D.Hey),	Borthwick Paper 102, University of York, 2002 (34pp)
18	The Making of Huddersfield,	Wharncliffe Books, 2003 (176pp) Illustrated
19	Christian Names in Local and Family History,	The National Archives, 2004 (190pp) Illustrated
20	Names and History,	Hambledon & London, 2004 (258pp) Illustrated

21   Places of Kirkheaton and   D. Shore, Huddersfield 2006
     District,   (56pp) Illustrated

22   The Place-names of   D. Shore, Huddersfield 2008
     Huddersfield,   (176pp) Illustrated

The following articles all appeared in the journal Old West Riding (1981-1993).
It was published twice annually 1981-87 inc., and annually 1988-95

23   'West Riding Emigrants 1843', OWR 1/1, 1981 (3pp)

24   'Steaners and Weirs', OWR 1/1, 1981 (5pp)

25   'The Origins of Yorkshire "royd" Surnames', OWR 1/1, 1981 (6pp)

26   'Tong Street', OWR 1/2, 1981 (4pp)

27   'Dog Pits', OWR 2/1, 1982 (1p)

28   'Colne or Holme'? OWR 2/2, 1982 (2pp)

29   'Self-help in House-building, 1799' (with C. Pearce), OWR 2/2, 1982 (3pp)

30   'Spring Woods', OWR 3/2, 1983 (6pp)

31   'Surnames and Settlement', OWR 4/2, 1984 (4pp)

32   'A Voyage from the West Indies', OWR 6/1, 1986 (3pp)

33   'Turf Pits', OWR 7/1, 1987 (2pp)

34   'Amer: a rare personal name', OWR 9, 1989 (1p)

35   'A View of the Moors', OWR 9, 1989 (1p)

36   'Hedges and Walls in West Yorkshire', OWR 10, 1990 (7pp)

37   'Stubbin', OWR 13, 1993 (1p)

Additionally,
38   'The Origins and Distribution of the Yorkshire "royd" Surnames',
     The University of Leeds Review, Vol. IX, No. 4, 1965

39   'Noms d'origine française dans le Yorkshire',
     Vie et Langage, No. 235, 1971

40   'Surnames and Place-names', The Local Historian, Vol. 10, No. 1, 1972

41   'Surname Heredity', The Local Historian, Vol. 10, No. 4, 1972

42   'Problems in the Identification of some Yorkshire Filial Names',
     Genealogists' Magazine, Vol. 17, No. 4, 1972

43   'Lancashire Surnames in Yorkshire',
     Genealogists' Magazine, Vol. 18, No. 1, 1975

44   'Migration and the Linguistic Development of Surnames',
     Family History, Vol. 11, Nos 77/78, New Series, Nos 53/54, 1980

45   'Personal Names and Surnames in some West Yorkshire "royds"',
     Nomina, Vol. 9, 1985

46 'A Moment in the History of Bradford Moor',
<u>The Bradford Antiquary</u>, Third Series, No. 1, 1985

47 'Shetcliffe in North Bierley',
<u>The Bradford Antiquary</u>, Third Series, No. 2, 1986

48 'Episodes in Lindley's History',
<u>Lindley Liberal Club Centenary 1887-1987</u>, 1987

49 'Tordoff, Tordiff, Torday – A Surname Study',
<u>GRINZ Year Book</u>, 1989 (N.Z.)

50 'Settlement in Huddersfield before 1800',
<u>Huddersfield: a Most Handsome Town</u>, E.A.H. Haigh (ed.),
Kirklees Cultural Services, 1992

51 'Clitheroe Wood and Farm, Almondbury, Huddersfield',
<u>Lancashire Local Historian</u>, No. 13, 1998

52 'The Opening up of Scammonden, a Pennine Moorland Valley' (with D.
Hey), <u>Landscapes</u>, Vol. 2, No. 1, Spring 2001 Illustrated

53 'Name of the Game', <u>History Today</u>, Vol. 54, Sept 2004

54 'The Sagars of Cliviger', <u>Lancashire Local Historian</u>, No. 18, 2005

55 'Some Regional Characteristics of Yorkshire Surnames',
<u>Yorkshire Names and Dialects</u>, Margaret Atherden (ed.), 2006

56 'Personal Names from 1300 to the Dissolution', and 'Surnames in the
Durham Liber Vitae after 1300' (with P. McClure), D. and L. Rollason
(eds), <u>The Durham Liber Vitae</u>, (2007)

Articles in <u>Ancestors</u>, the Family History Magazine of the National Archives:

	Date	Issue No.	Article Title
57	Feb./Mar. 2002	6	'All in the Genes'
58	Aug./Sept. 2002	9	'The Hallases: a Coal-mining family'
59	Dec. 2003/Jan. 2004	17	'Scenes from Ordinary Life'
60	April 2004	20	'The Name Game'
61	May 2004	21	'From First to Last'
62	June 2004	22	'Tom, Dick and Harry'
63	July 2004	23	'The History of Joseph'
64	August 2004	24	'Ranking Order'
65	Sept. 2004	25	'Upwardly Mobile'
66	Oct. 2004	26	'And the Saints Go Marching On'
67	Nov. 2004	27	'Scriptural Additions'
68	Dec. 2004	28	'Virgin Saints'
69	June 2005	34	'What's in a Name'

70	July 2005	35	'What are a Poet's Words Worth'?
71	Oct. 2005	38	'Nelson - an Admirable Name'
72	April 2006	44	'A Known Northern Name – Kneeshaw'
73	May 2006	45	'From Norfolk to Yorkshire – Camplejohn'
74	June 2006	46	'Agricultural Antecedents – Geldart'

Articles in <u>Nexus</u>, the New England Historic Genealogical Society bi-monthly magazine. I was a contributing editor for some years.

75	April 1988	Vol. V, No. 2	Emmeson/Em(p)son - on Emigration
76	June 1988	Vol. V, No. 3	English Origins – Hemingway
77	Oct. 1988	Vol. V, No. 5	English Origins – The Cordingleys and Taylors of Yorkshire
78	Dec. 1988	Vol. V, No. 6	Letters to the Editor – Cornelius (Empson)
79	Jan. 1989	Vol. VI, No. 1	English Origins – Banished from the Kingdom
80	June 1989	Vol. VI, Nos 3 & 4	English Origins – Thomas Wigglesworth of Pennsylvania
81	Oct. 1992	Vol. IX, No. 5	English Origins – 'Alias' Surnames (Part 1)
82	Feb. 1993	Vol. X, No. 1	English Origins – 'Alias' Surnames (Part 2)
83	May 1997	Vol. XIV, Nos 3 & 4	Waifs and Strays – A 'Peculiar' Find

A short series on dialect words featured in the <u>Yorkshire History Quarterly</u> (2001): Vol. 6, No. 4: Vol. 7, No. 1: Vol. 7, No. 2: Vol. 7, No. 3

Finally, between 1971 and 2000 I wrote well over 400 newspaper articles, mostly for the Huddersfield Examiner. Anybody who intends referring to them should know that they contain some errors, usually the fault of the newspaper but sometimes my responsibility. In the latter case I have usually had the opportunity to make corrections in a later publication. What follows is a résumé:

1) In the early 1970s The H.E. published 175 articles on Huddersfield surnames. A selection of these was then put together in the booklet <u>Surnames around Huddersfield</u>, listed above. Although these are now of limited value, since the publication of <u>Huddersfield and District Surnames</u> in the Yorkshire Surnames Series, they generally contain more examples and genealogical material.

2) Similarly, in 1977-79 the Bradford Telegraph and Argus published 51 articles on Bradford surnames. In this case also the publication of <u>Bradford and District Surnames</u> in the Yorkshire Surnames Series diminishes their value.

3) The H.E. series <u>Places and People</u> consists of 57 articles published between 29/7/76 and 1/9/77. A number of these were worked up from unpublished sources and have never found their way into later publications.

4) The first <u>In Focus</u> series consisted of 25 articles and it ran in the H.E. from 14/11/78 to 28/8/79. In the articles I took a fresh look at farms and hamlets scattered across the Huddersfield area.

5) There were 21 articles in the series <u>Know Your Town</u>. They were published from 3/2/80 to 21/6/80. They formed the basis of <u>Old Huddersfield: 1500-1800</u> (1981). I updated some of the subject matter in <u>Huddersfield: a Most Handsome Town</u> (1992) and <u>Names and History</u> (2004). <u>Place-names of Huddersfield</u> (2008) covers much new ground.

6) The second <u>In Focus</u> series followed the pattern established in the first series. There were 21 articles between 5/8/81 and 30/12/81.

7) The third <u>In Focus</u> series of 21 articles ran in 1982-83.

8) A series of 20 articles, under the heading <u>In Court</u>, was based on original material drawn mostly from the Quarter Sessions Rolls. It ran from 3/9/83 to 16/6/84 and culminated in the publication <u>Huddersfield and District under the Stuarts</u> (1985).

9) <u>Historic Landscapes</u> was a well-illustrated series of 19 articles that ran from 5/8/95 to 13/9/97.

Of several one-off articles, the only ones not covered by later publications are:
a) John Turner of Kirkburton – June 1980. Exact date unknown.
b) The Battling Battyes – 1984. Exact date unknown.
c) William Black of Huddersfield and Novia Scotia, 11/9/2000.

INDEX

This index draws attention to the place-names and surnames found in the text and the picture captions. Principal entries are not indexed. Surnames are identified with the symbol (s) and public houses are in italics. There may be more than one reference on a page.

Talbot 16
Tavenor(*s*) 153
Taylor(*s*) 79, 101
Teapot Chapel 5
Tel Aviv 43
Temple Street 86, 95, 125
Tetley(*s*) 146
The Flying Circus 116
The Grove 128
The Melting Pot 111
Thewlis(*s*) 120
Thornhill(*s*) 45, 53, 76, 88
Thornhill Cottage 44, 45
Thornhill Road 76
Thornton(*s*) 55, 87, 95, 109, 137
Thornton Lodge 18, 128
Thorpe Hall 69
Threadneedle Street 113
Tinker(*s*) 113
Tinker Croft 23, 67
Tolson(*s*) 133
Tolson's Yard 133
Tommy Clay Clough 45
Toothill 124
Top of Huddersfield, Top of
(the) town 65, 68, 89, 95, 108, 126, 137, 139, 145, 156
Town Avenue 43
Town brook 57, 58, 66, 68-70, 85, 138, 154
Town Crescent 43
Town ditch 138
Town end 138
Townend(*s*) 139
Town fields 83
Town gate 10, 12, 82, 97, 100, 104, 108, 116, 126, 138, 139
Town Hall 22, 109, 117
Town Head 126, 138
Town ings 10, 69, 85, 86, 118, 138, 139
Town lane 7, 59, 85, 138, 156
Town Place 43
Town street 7, 10, 40, 82, 89, 100, 125, 138, 139
Town Terrace 43

Town well 138
Towser Castle 12, 40
Trafalgar Mills 85
Trinity House 140
Trinity Street 55, 60, 63, 68, 105, 111, 126, 127, 139, 140, 151, 152, 155
Tunnel Street 30, 131
Turnbridge 43, 54, 112, 142-143, 149

Union St 12, 35, 68, 110, 157
Unna Way 83
Upper George Street 63
Upper Green 108
Upperhead Row 12, 62, 63, 65, 87, 88, 92, 120, 130, 131, 133, 138

Vagrant Office 62, 64, 127
Venn Street 125
Viaduct Street 147
Vicarage 20, 63, 65, 119, 146
Victoria Buildings 19
Victoria Inn 146
Victoria Lane 19, 75, 87, 92
Victoria Street 19, 110
Victoria Terrace 45
Violet Street 144

Wade(*s*) 47
Wakefield 6, 10, 12, 33, 38, 74, 83, 101, 139
Wakefield/Austerlands Road 116, 133
Walker(*s*) 136, 148, 153
Wappy Nick 93
Wappy Spring 148
Warrenside 118
Wasps' Nest 16
Watergate 12, 40, 112
Waterloo 148
Waterworks Building 63, 130, 149
Watson(*s*) 112
Waverley House 150
Wear(*s*) 65
Webster(*s*), 59

Well 137
Wells Mill 62
Wentworth(*s*) 53
Wentworth Street 141, 150
Westbourne Road 94, 95
Westgate 12, 35, 42, 43, 66, 68, 78, 84, 89, 95, 99, 100, 110, 113, 116, 120, 126, 132, 136, 137, 139
West Parade 66, 127, 139
West Place 151
Wharf Inn 12, 22
Whitacre, Whitaker(*s*) 53, 152, 153, 155
Whitacre Street 53
White Bear/Boar 16
White Hart 16
White Horse 20, 88
White Lion 16
White Swan 16, 136
Whitley(*s*) 28, 138
Whitley Hall 28, 29, 66
Wiggan Lane 124
Wilby(*s*) 157
Wilkinson(*s*) 45, 65, 94, 148
Wilks, Wilks Yard 68
Williamson (*s*) 34, 93
Willow Lane 70, 124
Wilson(*s*) 17, 125, 128
Windsor Court 17
Wood(*s*) 136
Woodhead (*s*) 97, 122
Woodhead 37, 62, 100
Woodhouse 34, 48, 53, 76, 85, 87
Woodman 33, 83
Wood Street 44
Wool Pack 112
Wormald(*s*) 156
Wrigley(*s*) 128

Yards 17
York 11, 33, 40
York House 24

Zetland Hotel 119
Zetland Street 77, 122